Lecture Notes in Computer Science 756

Edited by G. Goos and J. Hartmanis

Advisory Board: W. Brauer   D. Gries   J. Stoer

Josef Pieprzyk   Babak Sadeghiyan

Design of Hashing Algorithms

Springer-Verlag

Berlin Heidelberg New York
London Paris Tokyo
Hong Kong Barcelona
Budapest

Series Editors

Gerhard Goos
Universität Karlsruhe
Postfach 69 80
Vincenz-Priessnitz-Straße 1
D-76131 Karlsruhe, Germany

Juris Hartmanis
Cornell University
Department of Computer Science
4130 Upson Hall
Ithaca, NY 14853, USA

Authors

Josef Pieprzyk
Department of Computer Science, Centre for Computer Security Research
University of Wollongong
Wollongong, N.S.W. 2500, Australia

Babak Sadeghiyan
Computer Engineering Department, Amir-Kabir University of Technology
Tehran, Iran

CR Subject Classification (1991): E.3-4, G.2.1, F.2.2, D.4.6, C.2.0

ISBN 3-540-57500-6 Springer-Verlag Berlin Heidelberg New York
ISBN 0-387-57500-6 Springer-Verlag New York Berlin Heidelberg

Typesetting: Camera-ready by author
Printing and binding: Druckhaus Beltz, Hemsbach/Bergstr.
45/3140-543210 - Printed on acid-free paper

Preface

Historically, computer security is related to both cryptography and access control in operating systems. Cryptography, although mostly applied in the military and diplomacy, was used to protect communication channels and storage facilities (especially the backups). In the seventies there was a breakthrough in cryptography - the invention of public-key cryptography. It started in 1976 when Diffie and Hellman formulated their public-key distribution system and formally defined public-key cryptosystems. Two years later two practical implementations of public-key cryptosystems were published. One was designed by Rivest, Shamir, and Adleman (called the RSA system); the authors based the system on the two "difficult" numerical problems: discrete logarithm and factorization. The other invented by Merkle and Hellman was based on the knapsack problem, which is even "harder" than these used in the RSA system. Since that time cryptography, traditionally seen as the theory of encryption algorithms, has extended its scope enormously. Now it comprises many new areas, namely authentication, digital signature, hashing, secret sharing, design and verification of cryptographic protocols, zero knowledge protocols, quantum cryptography, etc.

This work presents recent developments in secure hashing algorithm design. The main part of the work was written when the authors were with the Department of Computer Science, University of New South Wales, Australian Defence Force Academy, and Babak Sadeghiyan was a PhD student working with Josef Pieprzyk as his supervisor.

Hashing is a process of creating a short digest (i.e. 64 bits) for a message of arbitrary length, for example 20 Mbytes. Hashing algorithms were first used for searching records in databases. These algorithms are designed to create a uniform distribution of collisions (two messages collide if their digests are the same). In cryptographic applications, hashing algorithms should be "collision-free", i.e. finding two different messages hashed to the same digest should be computationally intractable. Hashing algorithms are central for digital signature applications and are used for authentication without secrecy.

There have been many proposals for secure hash algorithms, and some of them have been in use for a while. However, many of them have proved insecure. One of the major reasons for this is the progress in technology. The failed effort of many researchers suggests that we should work on some guidelines or principles for the design of hash functions. This work presents some principles for the design of secure hash algorithms. Hash algorithms are classified based on whether they apply a block cipher as the underlying one-way function or not.

For a block-cipher-based hash scheme, if the underlying block cipher is secure against chosen plaintext/ciphertext attack, the hash scheme is secure against meet-in-the-middle attack. We develop structures, based on DES-like permutations and assuming the existence of pseudorandom function generators, which can be adopted both for the structure of block-cipher-based hash schemes and for the underlying block ciphers to be used in such schemes.

Non-block-cipher-based hash functions include a spectrum of many different proposals based on one-way functions from different branches of mathematics. So, in the book, generalized schemes for the construction of hash functions are developed, assuming the existence of a one-way permutation. The generalized constructions are improvements upon the Zheng, Matsumoto and Imai's hashing scheme, based on the duality between pseudorandom bit generators and hash functions, but they incorporate strong one-way permutations. It is shown that we can build such strong permutations with a three-layer construction, in a theoretical approach. Two schemes for the construction of families of strong one-way permutations are also proposed.

Acknowledgement

We were very fortunate to receive help from many people throughout the time of this project. Firstly, we would like to express our deep gratitude to Professor Jennifer Seberry for her critical comments, helpful suggestions and encouragement. Also we would like to thank Professor Tsutomu Matsumoto and Dr Rei Safavi-Naini for their thoughtful criticism, suggestions and corrections. We also received helpful comments about the work from Dr Lawrence Brown, Professor Andrzej Gościnski, Dr Thomas Hardjono, Dr Xian-Mo Zhang and Dr Yuliang Zheng. We thank all our friends from the Department of Computer Science, University College, University of NSW, for their friendliness and everyday support. In particular our thanks go to Dr George Gerrity, Mr Per Hoff, Mr Jeff Howard, Dr Jadwiga Indulska, Mr Martin Jaatun, Mr Ken Miles, Mr Andy Quaine and Mr Wen Ung. Finally we would like to thank Mrs Nilay Genctruck for proof-reading the final manuscript.

   This project was partially supported by the Australian Research Council grant number A49131885.

September 1993   Josef Pieprzyk

                                                      Babak Sadeghiyan

Contents

1 Introduction   **1**

   1.1   Background and Aims   1

       1.1.1   Introductory Comments   1

       1.1.2   Discussion of Public-key and Private-key Cryptography   2

       1.1.3   Digital Signature   5

       1.1.4   RSA Cryptosystem and Digital Signature   9

       1.1.5   Signature-Hashing Scheme   10

       1.1.6   Other Applications of Hash Functions   13

   1.2   Contents of the Book   14

2 Overview of Hash Functions   **18**

   2.1   Introduction   18

   2.2   Properties of Secure Hash Functions   19

   2.3   Definitions   20

       2.3.1   Strong and Weak Hash Functions   20

       2.3.2   Message Authentication Codes and Manipulation Detection Codes   22

       2.3.3   Block-cipher-based and Non-block-cipher-based Hash Functions   23

   2.4   Block-cipher-based Hash Functions   24

2.4.1   Rabin's Scheme   25

2.4.2   Cipher Block Chaining Scheme   26

2.4.3   CBC with Checksum Scheme   26

2.4.4   Combined Plaintext-Ciphertext Chaining Scheme . . .   27

2.4.5   Key Chaining Scheme   28

2.4.6   Winternitz' Key Chaining Schemes   29

2.4.7   Quisquater and Girault's 2n-bit Hash Function   30

2.4.8   Merkle's Scheme   31

2.4.9   N-hash Algorithm   32

2.4.10 MDC2 and MDC4   33

2.5   Non-block-cipher-based Hash Functions   34

2.5.1   Cipher Block Chaining with RSA   35

2.5.2   Schemes Based on Squaring   36

2.5.3   Schemes Based on Claw-Free Permutations   38

2.5.4   Schemes Based on the Knapsack Problem   39

2.5.5   Schemes Based on Cellular Automata   40

2.5.6   Software Hash Schemes   41

2.5.7   Matrix Hashing   43

2.5.8   Schnorr's FFT Hashing Scheme   44

2.6   Design Principles for Hash Functions   45

2.6.1   Serial Method   45

2.6.2   Parallel Method   46

2.7   Conclusions   46

3 Methods of Attack on Hash Functions   48

3.1   Introduction   48

3.2 General Attacks 49

3.3 Special Attacks 50

    3.3.1 Meet-in-the-middle Attack 51

    3.3.2 Generalized Meet-in-the-middle Attack 52

    3.3.3 Correcting Block Attack 53

    3.3.4 Attacks Depending on Algorithm Weaknesses 53

    3.3.5 Differential Cryptanalysis 54

3.4 Conclusions 54

4 Pseudorandomness   **56**

4.1 Introduction 56

4.2 Notation 58

4.3 Indistinguishability 58

4.4 Pseudorandom Bit Generators 60

4.5 Pseudorandom Function Generators 62

4.6 Pseudorandom Permutation Generators 66

    4.6.1 Construction 66

    4.6.2 Improvements and Implications 69

    4.6.3 Security 72

4.7 Conclusions 76

5 Construction of Super-Pseudorandom Permutations   **77**

5.1 Introduction 77

5.2 Super-Pseudorandom Permutations 78

5.3 Necessary and Sufficient Conditions 79

5.4 Super-Pseudorandomness in Generalized DES-like Permutations 92

    5.4.1 Feistel-Type Transformations 93

5.4.2 Super-Pseudorandomness of Type-1 Transformations . 96

5.5 Conclusions and Open Problems 103

6 A Sound Structure   105

6.1 Introduction 105

6.2 Preliminaries 106

6.3 Perfect Randomizers 112

6.4 A Construction for Super-Pseudorandom Permutation Generators 116

6.4.1 Super-Pseudorandomness of $\psi(h, 1, f, h, 1, f)$ 117

6.4.2 Super-Pseudorandomness of $\psi(f^2, 1, f, f^2, 1, f)$ 124

6.5 Conclusions and Open Problems 130

7 A Construction for One Way Hash Functions and Pseudorandom Bit Generators   132

7.1 Introduction 132

7.2 Notation 134

7.3 Preliminaries 135

7.4 Theoretic Constructions 137

7.4.1 Naor and Yung's Scheme 138

7.4.2 Zheng, Matsumoto and Imai's First Scheme 138

7.4.3 De Santis and Yung's Schemes 139

7.4.4 Rompel's Scheme 140

7.5 Hard Bits and Pseudorandom Bit Generation 140

7.6 A Strong One-Way Permutation 146

7.7 UOWHF Construction and PBG 151

7.7.1 UOWHF Based on the Strong One-way Permutation . 152

7.7.2   Parameterization 153

7.7.3   Compressing Arbitrary Length Messages 154

7.8   A Single construction for UOWHF and PBG 155

7.9   Conclusions and Extensions 156

8 How to Construct a Family of Strong One Way Permutations 157

8.1   Introduction 157

8.2   Preliminary Comments 159

8.3   Strong One Way Permutations 162

8.3.1   A Scheme for the Construction of Strong Permutations 164

8.3.2   A Three-layer Construction for Strong Permutations . 166

8.4   Conclusions 168

9 Conclusions   **170**

9.1   Summary 170

9.2   Limitations and Assumptions of the Results 174

9.3   Prospects for Further Research 177

Bibliography   **179**

Index   **191**

List of Symbols

C	Ciphertext
M	Message or a plaintext
K	Key
Σ	Alphabet
$\sum$	Summation
$\subset$	Subset
$\rightarrow$	Mapping
$\circ$	Composition of functions
$\in$	Set membership
$\equiv$	Congruence
$!$	Factorial
$\mid$	Such that (set notation)
$\oplus$	Exclusive-or (of Booleans)
$\vee$	Or (of Booleans)
$\wedge$	And (of Booleans)
$\parallel$	Concatenation
$O(f)$	Big-Oh of the function f
$\cup$	Union
$\lceil x \rceil$	Smallest integer greater than x
$\lfloor x \rfloor$	Greatest integer smaller than x
N	The set of natural numbers
Z_n	The set of integers modulo n
$\mid x \mid$	Absolute value of the number x
$\#S$	Number of elements in the set S
$\log_n$	Logarithm to the base n

Chapter 1

Introduction

1.1   Background and Aims

1.1.1   Introductory Comments

The development of telecommunication and computer technologies have brought us into an era in which inexpensive contact between people or computers on opposite sides of the world is commonplace. The existing services such as electronic mail, electronic funds transfer, and home banking have already changed our way of life. Electronic mail systems significantly reduce our reliance on paper as the major medium for exchange of information, by providing rapid and economic ways for the distribution of data. It is clear that, with the widespread implementation and use of such services, senders and receivers of sensitive or valuable information will require secure means for validating and authenticating the electronic messages they exchange. The least that may be expected of these services is that they should offer the same security level as that of the conventional mechanisms. In the mail service, conventional paper mail has its own envelope, which protects the secrecy of its contents, it is also signed which assures the recipient of its origin. Similar properties should have the electronic mail service.

On the other hand, the increased use of satellite, microwave, cellular mobile and other forms of radio communication allow the illicit interception of communications. Moreover, the widespread use of computers provides the interceptors with computer data, which can easily be edited to sort the valuable information. Figure 1.1 schematically shows such a scenario.

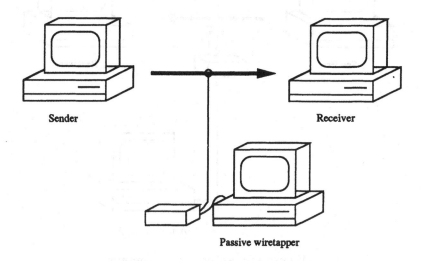

Figure 1.1: Passive Eavesdropping over Communication Networks

While eavesdropping on radio communications is a passive act, an active wiretapper can inject fraudulent messages in other types of communication links such as telephone networks. Figure 1.2 illustrates a possible active wiretapping scenario.

1.1.2   Discussion of Public-key and Private-key Cryptography

Cryptography is the study of mathematical systems for solving two kinds of security problems: privacy and authentication [Diffie and Hellman, 1976]. A privacy system prevents the extraction of information by unauthorized parties from messages transmitted over a public channel, thus assuring the sender of the message it will be read only by the intended recipient. An authentication system prevents the unauthorized injection of messages into a public channel, assuring the receiver of a message of its legitimacy. The authentication problem can be divided into *message authentication*, where the problem is assuring the receiver that the text has not changed since it left the sender, and *user authentication*, where the problem is verifying that

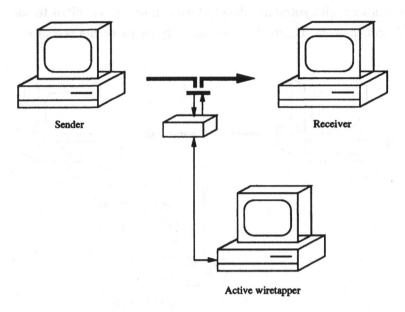

Figure 1.2: Active Eavesdropping

an individual is who they claim to be.

Once, cryptography was of interest only in military and diplomatic world. Now, with services such as electronic mail, and the existence of huge databases containing sensitive medical and personal data, the need for cryptography is evermore widespread in our society.

Encryption is a mathematical transformation or function applied to the message such that an eavesdropper is not able to extract any useful information about the original message from the transformed message (the transformed message is also called a cryptogram or a ciphertext). Along with the technological developments in teleprocessing, which have given rise to new secrecy and authenticity requirements, cryptographers have developed new encryption algorithms using complex mathematical systems [Moore, 1988]. The conventional cryptographic schemes are *single-key* or *private-key cryptosystems* where the transformation is controlled by a secret key. With new advances in computer technology, many conventional cryptographic schemes were eventually broken. Nevertheless, new and more complex private-key encryption schemes for the security of files were designed.

New families of cryptosystems known as *substitution-permutation networks* were developed, based on the theoretical works of Shannon ([Shannon, 1949b], [Shannon, 1949a]). They led to the development of systems such as Lucifer in 1973, DES in 1977, FEAL in 1987 [Shimizu and Miyaguchi, 1987], and LOKI in 1990 [Brown, 1991]. It is mentioned in [Brown, 1991] that:

> *Traditional cryptographic schemes have relied on a series of* substitution, *where letters or words are replaced by others, and* transposition *or* permutation, *where the order of letters or words is changed, operations to conceal the message. The particular substitution or transposition used is controlled by a* key. *This key is used by both the sender and the recipient of the concealed message, and hence has to be kept secret in order to protect the secrecy of the message. Such schemes are called* Private-key Cryptosystems *for this reason.*

These systems need to exchange keys via a private or secure channel, for example a trusted courier, to keep them secret. Figure 1.3 shows an application of a private-key cryptosystem to provide secrecy of transmitted messages via encryption.

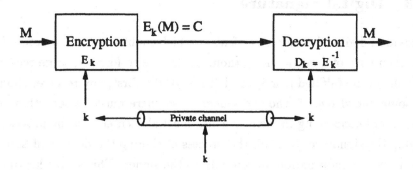

Figure 1.3: Traditional Cryptosystem with the Decryption Key Transferred over a Private Channel

Merkle suggests the reason why the private channel is not used for normal communication is because of its expense and inconvenience [1] [Merkle, 1978].

---

[1] Diffie and Hellman mentioned that:

A major problem with applications of private-key cryptosystems in large computer networks is key distribution, which require $\frac{n(n-1)}{2}$ key exchanges for n parties, unless some form of trusted key distribution hierarchy is used.

Another problem is its failure in resolving authentication problems arising from the dishonesty of either the sender or receiver. In electronic mail, for some messages, a degree of privacy or secrecy is needed, while authenticity is a requirement for nearly all messages. The technique of authentication with private-key cryptosystems is seriously deficient, since both the sender and receiver must know a secret key. The sender uses the key to generate an authenticator, and the receiver uses it to check the authenticator. Having this key, the receiver can also generate forged authenticators and therefore, may forge messages appearing to come from the sender. Hence, though this kind of authenticator can protect both sender and receiver against third party enemies, it cannot protect one against fraud committed by the other [Davies, 1983]. As a solution to the dispute problems, Diffie and Hellman proposed the use of digital signature based on public-key cryptosystems [Diffie and Hellman, 1976].

1.1.3   Digital Signature

To provide a digital signature, as a feature which enables anyone to determine the authenticity of a document without being able to forge it, some requirements should be fulfilled [Davies and Price, 1980]. First, any receiver should have some knowledge of who the sender is, so there must be something on public record concerning the sender which must also enter into the process of verifying the signature. Second, the process of signing the document should use some secret information known only to the signer. This secret key must be somehow related to the public information. Third, the signature must depend in a complex way on every digit of the message so that it would be impossible to modify the message and leave its signature unchanged. This requirement implies that the size of the signature field must be big enough

---

The secure channel cannot be used to transmit the message itself, for reasons of capacity or delay. For example, the secure channel might be a weekly courier and the insecure channel be a telephone line.

so the search of all possible messages for a given signature is intractable.

While a person's handsigned signature is constant, a digital signature depends upon the message. It can be computed only by the sender of the message, on the basis of some private information (known to the sender only). Digital signatures allow authentication of messages by guaranteeing that firstly no one (except the sender) is able to produce the sender's signature and secondly the sender is not able to deny their signature for the message they sent. The receiver can verify that no one tampered with the message while it was on its way to him, and the sender is confident that the receiver will not be able to change even one bit of the message without altering the signature.

Diffie and Hellman in their seminal paper [Diffie and Hellman, 1976], mentioned that:

> *Widening applications of teleprocessing have given rise to a need for new types of cryptographic systems, which minimize the need for secure key distribution channels and supply the equivalent of a written signature.*

> *At the same time that communications and computation have given rise to new cryptographic problems, their offspring, information theory, and the theory of computation have begun to supply tools for the solution of important problems in classical cryptography.*

The search for secure cryptosystems with more convenient features is one of the main themes of cryptographic research. In the nineteen-twenties the *one-time pad* cryptosystem was invented and later was shown to be *unconditionally secure*, that is the system can resist any cryptanalytic attack no matter how much computation is allowed. One time pads require long keys and are therefore expensive in most applications. However, the security of most other cryptographic schemes is based on the computational difficulty of discovering the plaintext without the knowledge of the key. These cryptosystems are called *computationally secure* that is, the system is secure because of the intractability of the cryptanalysis. The complexity theory classifies problems into classes depending on their computational difficulty. In general, if a cryptanalysis problem is solvable in polynomial time with polynomial-size computing resources, the corresponding cryptographic system is considered to be broken. Such a cryptanalysis problem is said to belong to the class **P**. All computational problems for which there is no polynomial-time algorithm, are collectively called intractable (for more details about different classes of intractability refer to [Garey and Johnson, 1979]). Using intractable computational problems, Diffie and Hellman and also independently Merkle [Merkle, 1978] modified the concept of a private channel by introducing *the public-key distribution system* or the concept of the public channel. They noted that it is possible to design systems in which two parties communicate over a public channel and use publicly known algorithms to exchange a secret piece of information. Later this concept was further elaborated by Rivest, Shamir and Adleman [Rivest *et al.*, 1978] and Merkle and Hellman [Merkle and Hellman 1978]. As the result of their works *public-key cryptography* has been invented.

In public-key cryptosystems, the receiver of a message generates beforehand a public key e and a secret key d. The public key e is used with a publicly known algorithm E for encryption, while the secret key d is used with algorithm D, which is also publicly known, for decryption. Figure 1.4 depicts a secrecy system with a public-key cryptosystem.

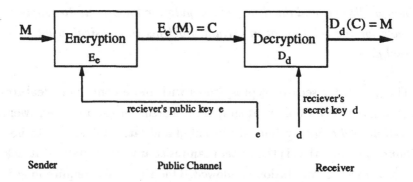

Figure 1.4: Public-key Cryptosystem with the Encryption Key Transferred over a Public Channel

To make a secure signature on the message, one can reverse the process of application of the two algorithms of the public-key cryptosystem. The sender of a message generates a public key e' and a secret key d'. The sender decrypts the message x with his secret key d' to get $S = D_{d'}(x)$ as his signature on x. A receiver can restore the message with the aid of the sender's public key e', by applying encryption as $E_{e'}(S)$. Figure 1.5 illustrates the mechanism of a digital signature based on a public-key cryptosystem.

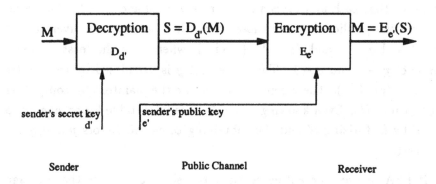

Figure 1.5: Principle of Digital Signature Based on a Public-key Cryptosystem

The restoration of the message text using the sender's public key verifies that S is coming from the sender as only he has access to the secret key and only he can generate the signature. To verify the signature, the receiver relies on the redundancy in the message in order to see that the result of applying the transformation E is a genuine message. If a wrong key was used, the result would be a random pattern with a high probability. If the message to be encrypted was random, there would be no way of resolving this dilemma. Merkle suggested that some controlled redundancy be deliberately introduced into the message so that it can be verified [Merkle, 1978].

It is noteworthy that not all public-key cryptosystems can be used for both privacy (secrecy) and authentication. Those that can, allow the process by which encryption and decryption is made to be reversed. The only known public-key cryptosystem which can be adapted for both authentication and secrecy was developed by Rivest, Shamir and Adleman, and is called the RSA cryptosystem [Rivest *et al.*, 1978].

1.1.4 RSA Cryptosystem and Digital Signature

In the RSA cryptosystem used for secrecy, the receiver chooses two large primes p and q and forms their product N. He keeps p and q secret, but makes N public. N must be large enough so that its factorization is infeasible. To meet this requirement, Davies suggested that N must be a number of 500 binary digits or more[2] [Davies, 1983]. A message is divided into blocks of a length such that each block x is a number between 0 and $N-1$. The secret key d and the public key e are chosen such that both are relatively prime to $p-1$ and $q-1$, and $de = 1 \pmod{\phi}$, where ϕ is the least common multiple of $p-1$ and $q-1$. The ciphertext y is calculated by the sender as $y = x^e \pmod{N}$. The receiver can recreate the plaintext by computing $x = y^d \pmod{N}$. Even knowing e and N, it is intractable for an enemy to derive d by factorizing N and thus obtaining ϕ, as the factors p and q are kept secret.

If RSA is used for authentication, the signature S for the message block x is calculated as $S = x^d \pmod{N}$ [Rivest *et al.*, 1978], [De Jonge and Chaum, 1986]. Although it is very elegant, there are some problems with the RSA signature scheme. Encryption is used mainly during the communication process, and the ciphertext is forgotten after the communication has been completed and the message has been recovered. However, a signature is kept for lifetime of the message, and it accompanies the message until it is destroyed. Signing each block separately using the RSA transformation produces a signature with the same length as the message. This is an expensive and unsatisfactory solution as it needs a double space for storage and a double bandwidth for transmission. Moreover, the computation required for RSA encryption or decryption is time consuming. Davies noted that it takes many minutes for a microcomputer to encrypt one block of 512 bits, and most messages or documents which need signature contain many such blocks [Davies, 1983]. Signing individual blocks also has the disadvantage as blocks may be fraudulently interchanged.

---

[2]More recently [Silverman, 1991] suggested that

> For the time being, even with much faster computers, 120 *digits* promises to be the limit of practical factoring.

where *digits* means decimal digits.

Moore analyzed the multiple application of the RSA signature scheme and showed how to obtain a forged signature[3] from a collection of valid ones [Moore, 1988]. There are several ways to do this. Here we describe a simple version of the attack. Consider a message m which can be represented as the product of two other messages, $m = uv \pmod{N}$. If someone could obtain a signature on m from a signing party, he would be able to forge a signature on u or v without needing to know the secret key d of the signing party. Since $m^d = (uv)^d = u^d v^d \pmod{N}$, then

$$m^d u^e = u^d v^d u^e = v^d \pmod{N}$$

and

$$m^d v^e = u^d v^d v^e = u^d \pmod{N}$$

This attack relies on the fact that the mathematical function which RSA uses namely exponentiation modulo a composite number, preserves the multiplicative structure of the input.

1.1.5   Signature-Hashing Scheme

Although the notion of a digital signature is one of the most fascinating features of public-key cryptography, the proof of security of public-key systems relies on assumptions that the underlying computational problems are intractable. In the RSA system both the factorization and discrete logarithm problems are assumed to be intractable. In addition, a practical implementation of a signature scheme is often made very difficult by the complexity of the algorithm needed in the system [Damgard, 1987]. Furthermore, the homomorphic structure of the underlying mathematical function makes the signature schemes vulnerable to attacks as outlined in the previous subsection. as that of the previous subsection. To protect signatures against such attacks, users can sign only meaningful messages, but this places the burden of security entirely on the users. These problems suggest the strategy of applying some suitable transformation to the message before signing it, in order to strengthen the signature system by destroying any structure in the underlying public-key algorithm. Mostly a signature-hashing scheme is used where the method is to apply a one-way hash function to the message before

---

[3]It was first pointed out in [Davida, 1982] and in [Denning, 1984].

it is signed. The message is thus signed by computing $S = D(h(M))$, where h is the one-way hash function. The values that the hash function generate are effectively random numbers that depend on all the bits of the message. Although M may be of any finite length, $h(M)$ is usually of a predetermined size.

Additionally, a hashing scheme is able to improve the speed of a signature scheme. For example, combining the RSA signature scheme with a collision free hash function we get a scheme which is more efficient and much more secure. The hash function compensates for the computationally intensive nature of the RSA algorithm by providing a compressing technique such that the whole document is summarized or represented in a checksum. Then the digital signature is applied to the compressed version. Note that some messages may be shorter than a block, in these cases the hash scheme does not improve the efficiency of calculating the signatures, but most messages contain several blocks. The function h is defined such that $h(M)$ can be calculated from the message M with easily, but if only $h(M)$ is known, finding even one message M that will generate this value is "difficult". Moreover, calculating any other message M' that yields the same hash value, i.e., $h(M) = h(M')$, must be infeasible. The hash value is subject to the signature process of the RSA method with the secret key of the sender d. The corresponding public key e is used by the receiver to invert the transformation and restore the value $h(M)$. At the receiving end, the function h is applied to the received message M, and the two values of $h(M)$ are compared. The signature is considered genuine if the two values are equal. Figure 1.6 depicts this procedure.

It would therefore not be necessary to divide the text into blocks and apply the signature process to each block separately. Instead it is sufficient to form a one-way hash function of the entire message and apply the signature to this function.

In the above scenario, the function h must be public knowledge, since the receiver applies it to the message. Possibly the biggest threat is the modification of existing or known messages for which the signature is available so that the same value is generated. It is noteworthy that here we are referring to all messages, not only "meaningful" or "useful" messages. The strength of the above method depends essentially on the inability of a forger to construct

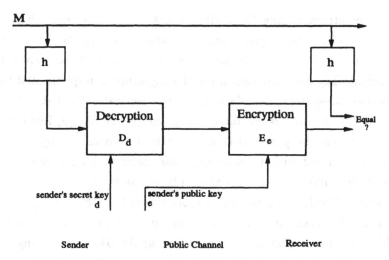

Figure 1.6: Digital Signature with the Application of a Hash Function

any message that matches a given hash value.

A hash function can be defined to be a cryptographic function for computing a fixed length digest of a message [Akl, 1983], [Preneel *et al.*, 1992]. Denning lists the following four properties that the function h should satisfy[4] [Denning, 1984]:

1. h should destroy all homomorphic structures in the underlying public-key cryptosystem.

2. h should be able to be computed over the entire message, rather than just on a block.

3. h should be one-way so that messages are not disclosed by their signatures.

4. h should have the property that for any given message M and value $h(M)$, it is computationally infeasible to find another message M' such that $h(M) = h(M')$.

---

[4]Hash functions, as functions for computing fixed length compressions of messages, can support operations such as INSERT, SEARCH and DELETE in computer systems. However, the requirements for such hash functions is looser than that of the hash functions required in cryptography.

Besides strengthening the digital signature, the hashing scheme provides several other advantages. First, it separates the signature transformation from the secrecy transformation, allowing secrecy to be implemented with a private-key cryptosystem while the signature is implemented with a public-key cryptosystem. An advantage of this separation is that in the context of the ISO Open System Interconnect Reference Model, integrity and confidentiality can be provided at different layers. Second, signatures can be publicly disclosed without revealing their corresponding messages. This is particularly important when recovering from compromises or disclosure of private keys. Third, it can provide a more efficient method of signing messages. The RSA transformation, for example, is several times slower than the DES. So it is considerably faster to first apply DES for hashing a long message down to a single block and then applying the RSA signature, than applying the RSA signature to the entire message.

1.1.6 Other Applications of Hash Functions

In the previous section, we described the use of hashing schemes to provide efficient and secure digital signatures in message handling systems. This is not the only application of hash functions in cryptography. For example, when the integrity of a file is to be protected against illicit alteration, the owner of the file can obtain a hash version of the file. Now, the file can be stored on a public medium. Whenever the file is to be used, its owner computes the hash value of the file and compares it with the stored copy. If they are equal, the file is intact and has not been tampered with. It is noteworthy that hash functions play an important role in the design of efficient cryptographic protocols [Preneel *et al.*, 1992].

In [Sadeghiyan, 1991] an overview of secure electronic mail has been presented, where the importance of secure hash functions in providing many different security services such as message integrity check, message origin authentication check, message authentication check, and key management is demonstrated.

The need for secure hash functions has been realized before [Denning, 1984] and [Davies and Price, 1980]. Several attempts have been made to construct such functions using encryption algorithms such as DES or RSA.

However, none of these schemes have proved to be secure, and several of these proposals using DES have also been proven to be insecure [Damgard, 1987]. As the security of many cryptographic services and schemes reduces to the existence of a secure hash function, the aim of this book is to discuss design rules for the construction of secure hash functions.

As many cryptographic hashing schemes rely on the application of block cryptosystems such as DES or LOKI, in the first part of this book we will develop a new construction for block cryptosystems in order that they can be applied in secure hashing schemes.

It has also been shown that the existence of a secure hash function depends on the existence of a one-way function, and in practice many schemes for hash functions take advantage of the application of such functions. In the second part of the book we consider the application of one-way functions in hashing schemes and we show how to construct a secure hash function given the existence of a one-way permutation. As one-way hash functions are, in a sense, duals of pseudorandom bit generators, we also show how to construct a module which can be used for the construction of both pseudorandom bit generators and one-way hash functions.

1.2   Contents of the Book

The book is arranged as follows. In Chapter 2 an overview of cryptographic hash functions is presented. The chapter shows how much effort has been put into the design of secure hash functions, and also demonstrates that the design of an efficient and provably secure hash function has been less successful. It also gives a division of hash functions based on whether the hash scheme uses a block cipher in their structure. First, the requirements of a "good" hash algorithm are described. Second, formal definitions of hash functions are presented. Third, the classification of CCITT for hash functions in secure message handling standards is stated. Then, several proposals for hash schemes are presented and known attacks on them noted. They are also categorized based on whether they incorporate a block cipher.

In Chapter 3, methods of attack on hash functions are presented. This chapter only describes some rather general methods of attack on hash functions and how these attacks work. First, an attack based on the *birthday paradox*, named after a problem in probability theory, is described. This attack is a general method of attack and can be launched against any hash scheme. Then a special version of it, known as the *meet-in-the-middle* attack, is presented; this can be launched against schemes which employ block chaining in their structure. The probability of success for the above mentioned attacks depends on the length of the hash value and how randomly the cryptographic algorithm performs. Some other methods of attack against hash functions are also described where specific weaknesses of the algorithms are exploited to find collisions.

In Chapter 4 recent developments in the theory of pseudorandomness based on a complexity theoretic approach are presented. A short discussion on pseudorandomness of a block cipher and its relation to the birthday attack and the meet-in-the-middle attack is also presented. First, concepts of pseudorandomness and indistinguishability are introduced and, based on that, definitions for pseudorandom bit generators and pseudorandom function generators are presented. Then distinguishing circuits are defined. Note that any chosen plaintext attack against a block cipher is equivalent to a distinguishing circuit. Later the construction of pseudorandom permutation generators is described using Luby and Rackoff structure with three rounds of DES-like permutations and with three pseudorandom functions. Luby abd Rackoff used this result to justify the application of DES-like permutations in the structure of DES. Then super-distinguishing circuits are introduced, where a chosen plaintext/ciphertext attack against a block cipher is equivalent to a super-distinguishing circuit. A meet-in-the-middle attack against a block-cipher-based hash scheme for finding colliding messages is virtually a chosen plaintext/ciphertext attack against the underlying block cipher. Furthermore, if there is a super distinguishing circuit for a block cryptosystem, there is a possibility of making collision messages for the corresponding block-cipher-based hash scheme. We conclude this chapter with the result that, if a block cipher is to be applied for hashing messages, it should be secure against a chosen plaintext/ciphertext attack. In other words, the cryptosystem family should be a super-pseudorandom permutation generator.

In Chapter 5, necessary and sufficient conditions for the construction of super-pseudorandom permutation generators are presented. After developing a convenient type for distinguishing circuits, we show the conditions that a DES-like construction for cryptosystems should satisfy, in order to achieve super-pseudorandomness. Based on this result we suggest that it is possible to achieve a super-pseudorandom permutation generator with four rounds of DES-like permutations and two independent pseudorandom functions. Next we generalize the above results and apply them for the generalized type DES-like permutations. We show how to construct a super-pseudorandom permutation using k^2 rounds of type-1 Feistel type transformations, where k is the number of branches of the structure.

In Chapter 6, we show how to construct a super-pseudorandom permutation generator from a single pseudorandom function generator. First, we develop a construction with two modules of Luby and Rackoff structure. We show that if two of the random functions are replaced by two random permutations, then each branch of the construction becomes independent of the other, and it is possible to make a perfect randomizer with two independent random functions. Then, based on the structure of this perfect randomizer, we show that it is possible to make another structure with only a single random function, where the two structures are indistinguishable from each other. The result of this chapter is that a new structure for the construction of block ciphers secure against a chosen plaintext/ciphertext attack is developed. This structure can be used in the design of block ciphers for hashing schemes.

The other class of hash schemes consists of those in which a one-way function other than a block cipher has been used. In Chapters 7 and 8 we consider the construction of hash functions based on one-way permutations. In Chapter 7 a construction for one-way hash functions and pseudorandom bit generators is presented. First, some definitions for one-way functions, hard bits of one-way functions and pseudorandom bit generators are given. Then some complexity-theoretic constructions for hash functions are reviewed. The Zheng-Matsumoto-Imai (ZMI) hashing scheme, as a dual of Blum-Micali pseudorandom bit generators is presented. To improve the efficiency of the ZMI scheme, we introduce the notion of strong one-way permutations. Next, given the existence of a one-way permutation, we introduce a method to make a strong one-way permutation, where calculating every bit of the input is as difficult as inverting the one-way permutation itself. Finally, we apply the proposed strong one-way permutation to construct a module which can be used for pseudorandom bit generation and secure hashing schemes.

In Chapter 8, we propose a practical way of constructing a family of strong one-way permutations. This family has the property that when a member is selected randomly, it is a strong one-way permutation. We use polynomials in a Galois field. Two methods are proposed. The first method is based on the composition of several rounds of a randomly chosen polynomial with any one-way permutation. The other method is based on a threefold composition, by applying a one-way permutation which we call a hiding permutation.

In Chapter 9, some concluding remarks are given.

Chapter 2

Overview of Hash Functions

2.1 Introduction

As we described in Chapter 1 the authentication of a message M is a procedure that allows two or more communicators to verify the authenticity of a document so that any fraudulent or accidental modification of the message is detected by the intended receiver. The techniques for authentication of messages are usually based on the redundancy contained in the message or are based on checks on some appropriate redundant information added to the message. The redundant information can be calculated as a hashing function for the message or it can be computed by an encryption algorithm for the message using a secret key known only to the communicating parties. It is important that a good authentication check be computed in such a way that the introduction of bogus messages into the communication network and the partial modification of genuine messages already present in the network is practically intractable.

In some cryptographic formats, each block of the plaintext message contains checksum bits that are appended to the block prior to encryption. The checksum is visible after decrypting the ciphertext. Notice that it is not the responsibility of the encryption protocol to protect communication against noise in the channel; that is for other communication layers to handle. The checksum is there to help determine whether the decryption was successful, in case the receiver selected the wrong key. This is particularly important when little or no error checking is used in the lower communication layers. The checksum alone cannot detect the work of a clever, active, wiretapper. That is what message digests are for.

In hashing-signature schemes, the signature is condensed by the use of a one-way hashing algorithm to form a small message digest of the entire message. The digest is analogous to a checksum, but it must be practically impossible to make another message that maps to the same digest or hash value. As an example of a bad choice of hashing algorithm, suppose that the Hamming error correcting code is used to form a check. An opponent could easily modify the message by inverting some information bits and those parity bits which are a function of the inverted information bits. The modified message would appear genuine to a receiver as the verifying procedure could not detect the modifications. Hence, for a hash scheme to be suitable for digital signatures some additional requirements should be satisfied.

2.2 Properties of Secure Hash Functions

When a hash function is applied to provide a secure hashing-signature scheme for electronic mail or documents, one important criterion is that the set of all hash values be nearly one-to-one with respect to the set of all message texts [Jueneman, 1987]. In other words, it is desirable that the checksums of two messages be identical if and only if those messages are identical. In general, this is impossible to satisfy when messages are longer than the hash value. However, Jueneman specifically lists the following properties for a secure hashing algorithm in [Jueneman, 1987]:

1. The hashing algorithm should be executed efficiently on computers with no need for special purpose cryptographic hardware.

2. The hash value must be sensitive to all possible permutations and rearrangements, as well as the edition, deletion, and insertion of the text.

3. If two different texts are compressed, the probability that the two hash values are equal should be a uniformly distributed random variable.

4. The length in bits of the hash value should be long enough so that it resists the so-called *birthday attack*. With today's technology this

value should be of the order of 128 bits. We will explain more about this attack later.

5. The hashing algorithm must not be invertible, nor subject to decomposition into separate and independent elements.

Concerning the last requirement, when the hashing algorithm is subject to decomposition into separate and independent elements, each element may be small enough that the *birthday attack* is feasible, from the computational time and storage point of view. Moreover, if the hashing algorithm were invertible, it would be possible to work both forward and backward to produce matching values. We will say more about matching by working forward and backward in Chapter 4, where we discuss the issue of *meet-in-the-middle* attacks.

2.3   Definitions

2.3.1   Strong and Weak Hash Functions

There have been many proposals for hashing algorithms, and they can be divided into two broad categories, based on their level of security: *collision-free hash functions* and *universal one-way hash functions*. Merkle uses the names *strong one-way hash function* and *weak one-way hash function* respectively [Merkle, 1979], [Merkle, 1989a], [Merkle, 1989b]. We use the terms interchangeably.

A strong one-way hash function or a collision-free hash function is a function h such that:

1. h can be applied to any message or document M of any size.

2. h produces a fixed size output.

3. Given h and M, it is easy to compute $h(M)$.

4. Given the description of the function h, it is computationally infeasible to find two distinct messages which hash to the same value[1].

---

[1] Messages which hash to the same value are called colliding messages or collisions.

On the other hand, a weak one-way hash function or a universal one-way hash function is a function that:

1. h can be applied to any message or document M of any size.

2. h produces a fixed size output.

3. Given h and M, it is easy to compute $h(M)$.

4. Given the description of the function h and a randomly chosen message M, it is computationally intractable to find another message which hashes to the same value.

Strong one-way hash functions are easier to use in systems than weak one-way hash functions, because there are no preconditions on the selection of the messages.

With weak one-way hash functions, there is no guarantee that finding a pair of messages which map to the same hash value is difficult. Thus, there may be messages m and m' that map onto the same hash value. However, deliberately picking a message equal to m or m' must be prevented. Furthermore, there should not be too many of those pairs; otherwise a randomly chosen message would not be safe. Thus, with a weak one-way hash function, finding another message which hashes to the same value as some randomly chosen m should be difficult. However, the message m may be chosen non-randomly, if the function h is random. Thus, many weak one-way hash functions have been described based on DES or on other good block ciphers [Merkle, 1979], [Merkle, 1989b].

In order to introduce randomness into the weak hashing algorithms, various methods have been proposed. One method is to randomize the message by encrypting it with a good block cipher using a truly random key. The random key would also be added at the start of the resulting ciphertext [Merkle, 1989a]. Another method is to select a random prefix to the message before running the hash algorithm. Such a random prefix would effectively randomize the hash value. Yet another method is to choose the hash function randomly from a family of hash functions instead of randomizing the message itself [Carter and Wegman, 1979].

Weak one-way hash functions are weakened when they are used repeatedly. As more messages are signed with the same weak one-way hash function, the chance of finding a message with a hash value equal to the hash value of a previous message increases. Hence, the overall security of the system is reduced. In contrast, strong one-way hash functions provide full security, even when applied repeatedly [Merkle, 1989b]. Stating that a hash scheme is *secure* usually means it is secure in the 'strong' sense, unless the context implies otherwise. In this chapter, when we say that a particular hash scheme is *secure*, we mean that there is yet no attack to find two distinct messages which hash to the same value (digest), with the computing resources that today's technology provides.

2.3.2   Message Authentication Codes and Manipulation Detection Codes

As the presence of redundancy in the message distinguishes authentic information from bogus information, in some methods the modification of information is detected through the distortion of the internal redundancy of the information. A universal algorithm, however, has to protect the integrity of the information without any assumption about the internal structure of the message [Preneel *et al.*, 1992]. There are two major approaches to introducing controlled redundancy into the information. The redundant information can be calculated as a cryptographic hash function of the message M under the control of a secret key K known only to the communicating entities. In this case, the redundant information is called Message Authentication Code or MAC. The ANSI9.9 message authentication standard represents one such technique. MACs make use of traditional cryptographic algorithms such as the DES, and rely on a secret authentication key to ensure that only authorized persons can generate a message with the appropriate MAC. As we mentioned earlier, RSA digital signature can be used to establish both the authenticity of a document and its origin. Because of the intensely computational nature of RSA, most digital signature schemes make use of hashing techniques. A MAC approach based on DES or other traditional cryptographic algorithms is often used for this purpose.

As we mentioned earlier, a secure MAC scheme should prevent even the owner of the private key from finding a collision message; otherwise the scheme would not be able to solve the problem of disputes arising between two communicating parties.

There are sometimes advantages in using a hashing algorithm which does not require a secret key. When the message is encrypted to provide confidentiality, it is preferable to provide the MAC on the plaintext instead of computing it on the ciphertext [Montolivo and Wolfowicz, 1987] so that authentication is independent of secrecy. Thus, authentication without confidentiality is possible, and even if the encryption scheme is broken, the authenticity of the information is still assured. The disadvantage of this method is the additional burden of the key management.

The other approach is computing the redundant information as a hash function of the message M alone, without requiring the use of a cryptographic key. In this case, the redundant information is called Manipulation Detection Code or MDC. As the hash function for producing the MDC is publicly known, the message together with the MDC is usually enciphered in order to prevent an attacker from succeeding in substituting his own MDC along with the modified text. The advantage of MDC is that only publicly known elements are required; as a result, it simplifies the key management in secure message-handling systems. In addition, as the authentication is separated from the encryption function or its mode of operation, encryption and message authentication can be implemented in different protocol layers in the context of the OSI reference model. The disadvantage of this approach is that, if the confidentiality mechanism is compromised, then there can be no assurance of the integrity.

2.3.3   Block-cipher-based and Non-block-cipher-based Hash Functions

One of the requirements of hash functions mentioned in [Denning, 1984] is that a hash function should be a one-way function, where a one-way function is a function which is easy to compute but difficult to invert. Rompel has shown that one-way functions are necessary and sufficient conditions for the construction of secure hash schemes [Rompel, 1990]. In other words the existence of a secure hash scheme depends on the existence of a one-way function. For a secure block cipher, given a ciphertext and the corresponding plaintext, it is difficult to find the key. Hence, many proposals for hashing algorithms have used a traditional block cipher as the underlying one-way function. Other proposals apply other types of functions which are considered to be one-way.

Not every block cipher algorithm may be suitable for the construction of a hashing algorithm. In Section 2.4 we present some proposals based on the application of known block ciphers such as DES. We also mention whether they have remained secure, or have been successfully attacked. In Chapter 3 we present methods of attack on hashing algorithms. Based on that, we explain in Chapter 4 what property a block cipher might have in order to be suitable for the construction of secure hashing algorithms.

There have been many proposals based on functions from number theory and other fields of mathematics which are considered to be one-way. In Section 2.5 we present some of these hashing functions and mention whether there has been any successful attack on them or not. In Chapter 8 we show how to make an efficient one-way hash function from any one-way permutation.

2.4   Block-cipher-based Hash Functions

To minimize the effort in the design of a cryptographically secure hash function, many designers of hash functions tend to base their schemes on existing encryption algorithms. In this section we present an overview of such schemes.

The general scheme for the construction of a hash function based on the application of a block cipher algorithm is to divide the message or the document into blocks. The blocklength is equal to the input or the key of the block cipher algorithm, depending on the scheme. If the length of the message is not a multiple of the blocklength, then the information is usually encoded and an additional block containing the binary representation of the added bits is appended to the message. To provide a randomizing element, an initial vector is normally used. This vector is denoted by IV and its value is either well known, or exchanged along with a key, or prefixed onto the message. The encryption algorithm E is denoted by

$$E(K, M)$$

where M is the input to the algorithm and K is the key. The proof of the security of such schemes relies on the collision freeness of the encryption algorithm used. We will return to this matter in section 2.6 where we explain Damgard's design principle [Damgard, 1987] and Merkle's meta method [Merkle, 1989b].

2.4.1 Rabin's Scheme

As an application of encryption algorithms in the construction of secure hash functions, consider the scheme proposed by Rabin in [Rabin, 1978]. Rabin's scheme can be described as follows. First the message is divided into blocks whose lengths are equal to the length of the input of the encryption algorithm. If the encryption algorithm is DES, for example, then the message is divided into blocks of 64 bits. Suppose that t blocks have resulted. Then the following computations are performed (see Figure 2.1)

$$
\begin{aligned}
H_0 &= IV \\
H_i &= E(M_i, H_{i-1}) \qquad i = 1, 2, \ldots, t \\
H(M) &= H_t
\end{aligned}
$$

Although Rabin's scheme is simple and elegant, Yuval demonstrated how it

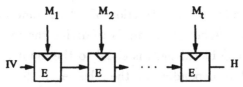

Figure 2.1: Rabin's Hashing Scheme

is susceptible to the so-called *birthday attack* when the size of the hash value is 64 bits [Yuval, 1979]. The scheme is also susceptible to *meet-in-the-middle attack*. We explain more about these attacks in Chapter 3.

2.4.2   Cipher Block Chaining Scheme

A widespread method for computing a hash value is the application of the
cipher block chaining mode of a block cipher algorithm. In this scheme,
the hash value is the last block of the ciphertext that resulted from the
application of the encryption algorithm in cipher block chaining (CBC) mode
to the message [DES, 1985], [DES, 1983], while the key and the initial value
are kept public. The scheme can be described as follows (see Figure 2.2)

$$
\begin{aligned}
H_0 &= IV \\
H_i &= E(K, M_i \oplus H_{i-1}) \qquad i = 1, 2, \ldots, t \\
H(M) &= H_t
\end{aligned}
$$

A variation of the above method is to apply the encryption algorithm in the

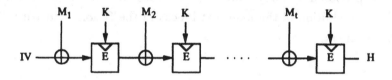

Figure 2.2: Cipher Block Chaining Scheme

cipher feedback (CFB) mode.

The security of this scheme depends on the error propagation properties
of the applied mode of operation.

2.4.3   CBC with Checksum Scheme

Another variation on the previous scheme is to add some redundancy to
the message in the form of the *exclusive-OR* of the plaintext blocks, where
the initial vector is assumed to be zero. The redundant information forms
a checksum which is appended to the plaintext blocks [Akl, 1983]. Subsequently, a block encryption algorithm in either cipher block chaining (CBC)
mode or cipher feedback (CFB) mode or output feedback (OFB) mode is
applied to the entire bit string [Meyer and Matyas, 1982]. In this scheme the

key can be either private or publicly known. If the key is secret, the security of the scheme depends on the mode of the encryption algorithm. However, the three proposed modes have been shown to be susceptible to attacks based on manipulations of blocks [Jueneman, 1982]. Table 2.1 is from [Preneel *et al.*, 1992] and indicates which manipulations are possible on blocks, so that the manipulations are not detectable. If the key is not secret, the scheme is also susceptible to *meet-in-the-middle* attack. To improve the above scheme,

mode	CBC	CFB	OFB
insertion	√	√	
permutation	√		√
substitution			√

Table 2.1: Manipulations on Blocks, in CBC, CFB and OFB Modes

a version was proposed by [Meyer and Matyas, 1982] where the checksum is provided by the addition of the plaintext blocks in the Galois field with 2^m elements for some m.

2.4.4 Combined Plaintext-Ciphertext Chaining Scheme

If we are applying a block cipher algorithm for both encryption of the message and generation of the hash value, different keys should be used for each operation, otherwise the scheme would be susceptible to several manipulations [Meyer and Matyas, 1982]. However, [Meyer and Matyas, 1982] proposed a scheme which needs one secret key to provide both secrecy and authentication. The description of the scheme is as follows (see Figure 2.3)

$$
\begin{aligned}
M &= M_t \dots M_1 \\
M_{t+1} &= IV \\
H_i &= E(K, M_i \oplus M_{i-1} \oplus H_{i-1}) \qquad i = 1, 2, \dots, t \\
H(M) &= H_{t+1}
\end{aligned}
$$

In the above scheme M_0 and H_0 are considered to be equal to zero. While $H(M)$ provides a hash of the message the H_i provide the ciphertext blocks. It is noteworthy that this algorithm is also susceptible to the *birthday attack*.

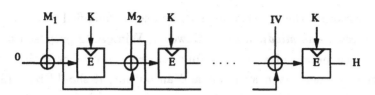

Figure 2.3: Meyer and Matyas's Combined Plaintext-Ciphertext Chaining Scheme

2.4.5 Key Chaining Scheme

This scheme has been proposed in [Davies, 1983] and [Denning, 1984] and is an improvement of Rabin's scheme. It can be described as follows.

$$
\begin{aligned}
H_0 &= IV \\
H_i &= E(M_i \oplus H_{i-1}, H_{i-1}) \qquad i = 1, 2, \ldots, t \\
H(M) &= H_t
\end{aligned}
$$

Although the scheme is an improvement of Rabin's scheme, it is still subject to *meet-in-the-middle attack*. Several modifications have been proposed to improve the scheme further. The first, proposed in [Davies and Price, 1980], is to repeat the scheme twice over the message, and the second modification is to execute the above algorithm with two different initial values. Coppersmith, however, showed that the *meet-in-the-middle* attack can still break these improved versions of the scheme [Coppersmith, 1985]. The third proposal is to first encrypt the message in CBC or CFB mode before applying the hash scheme. The fourth proposal is to append a checksum of all the message blocks before the execution of the hash scheme [Seberry and Pieprzyk, 1989].

When DES is used as the block cipher, each of the above schemes is vulnerable to attacks exploiting keys with some weaknesses. Quisquater and Delescaille worked on a collision search algorithm, which resulted in an attack on the fourth modified scheme [Quisquater and Delescaille, 1989a] and [Quisquater and Delescaille, 1989b].

2.4.6 Winternitz' Key Chaining Schemes

As we mentioned, the key chaining scheme and its modified versions are subject to *meet-in-the-middle* attack. However, Winternitz proposed a scheme for the construction of a one-way function from any block cipher. In any good block cipher, given an input and an output, it should be difficult to work out the applied key, while given the output and the key it should be easy to compute the input. In Winternitz' construction we are able to make a one-way function from any good block cipher so that, given the output and the key, it is difficult to guess at the value of the input. The construction is defined as

$$E^*(K, M) = E(K, M) \oplus M$$

Based on this construction, Donald Davies proposed a hash algorithm which can be described as in Figure 2.4.

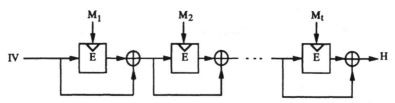

Figure 2.4: Davies' Scheme

$$H_0 = IV$$
$$H_i = E(M_i, H_{i-1}) \oplus H_{i-1} \qquad i = 1, 2, \ldots, t$$
$$H(M) = H_t$$

A similar scheme was proposed by Matyas, Meyer and Oseas and is described in Figure 2.5.

$$H_0 = IV$$
$$H_i = E(H_{i-1}, M_i) \oplus M_i \qquad i = 1, 2, \ldots, t$$
$$H(M) = H_t$$

Both of the above schemes were intended to be implemented with DES, and so under certain conditions the schemes are vulnerable to attacks based on

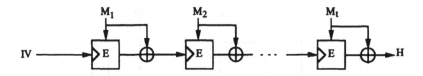

Figure 2.5: Meyer, Matyas and Oseas' Scheme

weak keys [Preneel *et al.*, 1992] or a key collision search [Quisquater and Delescaille, 1989b], while the threat of meet-in-the-middle attack has been thwarted because of the one-wayness of the applied function.

2.4.7   Quisquater and Girault's 2n-bit Hash Function

It is possible to attack all the hash schemes which produce 64 bit hash values with the *birthday* attack, since one need only obtain 2^{32} messages and their corresponding hash values to find collisions. As all of today's encryption schemes such as DES, FEAL and LOKI are 64-bit block ciphers, there have been many attempts to design schemes based on 64-bit block ciphers which result in a 128-bit hash value. One of the simplest solutions is to repeat a 64-bit scheme for two different values of a parameter, such as the initial value or the key.

One such attempt was made by Quisquater and Girault where they suggested a 128-bit hash algorithm using a 64-bit block cipher [Quisquater and Girault, 1989]. They took DES as the underlying block cipher. The description of their algorithm is given in Figure 2.6.

$$
\begin{aligned}
H1_0 &= IV_1 \\
H2_0 &= IV_2 \\
T1_i &= E(M1_i, H1_{i-1} \oplus M2_i) \oplus M2_i \\
T2_i &= E(M2_i, H2_{i-1} \oplus M1_i \oplus T1_i) \oplus M1_i \\
H1_i &= H1_{i-1} \oplus H2_{i-1} \oplus T2_i \\
H2_i &= H1_{i-1} \oplus H2_{i-1} \oplus T1_i \\
H(M) &= H1_t \parallel H2_t
\end{aligned}
$$

As DES is the underlying block cipher, some keys with certain weaknesses

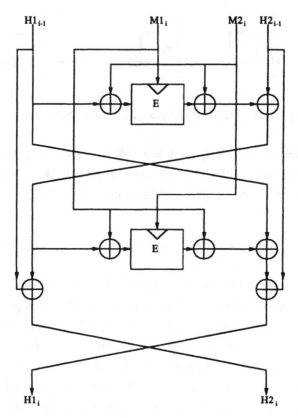

Figure 2.6: Quisquater and Girault's 2n-bit Scheme

can be exploited to compute collision messages for the scheme. In [Miyaguchi *et al.*, 1990], it has been shown how to make collision messages for the above scheme using the complementation property and the weak keys of DES. Moreover, [Preneel *et al.*, 1992] report that Coppersmith has broken this scheme for every block cipher because of linearities.

2.4.8 Merkle's Scheme

Based on Winternitz' construction, Merkle proposed several schemes in [Merkle, 1979], [Merkle, 1989a] and [Merkle, 1989b]. These schemes based on the application of DES, result in hash value of around 128 bits. The construction of these schemes follows a general method for the construction of hash algorithms. Merkle called it the *meta method*, which is the same as the serial method of design principles described by Damgard in [Damgard, 1989]. We describe the method later in this chapter, in Section 2.6. Merkle's proposals take advantage of the construction of Winternitz. In some proposals with complex and fast implementation, the message is first divided into blocks of 106 bits. The concatenation of each 106-bit block M_i of data with the 128-bit block H_{i-1} and the hash value result of the previous stage, makes X_i, a 234-bit block. We denote this concatenation by $X_i = M_i \parallel H_{i-1}$. Each block X_i is further divided into two pieces, X_{i1} and X_{i2} for each of the 117-bits in size. The description of the method is as follows (see Figure 2.7).

$$
\begin{aligned}
H_0 &= IV \\
X_i &= H_{i-1} \parallel M_i \\
H_i &= E^*(00 \parallel \text{first 59 bits of}\{E^*(100 \parallel X_{1i})\} \parallel \\
&\quad \text{first 59 bits of}\{E^*(101 \parallel X_{2i})\} \parallel \\
&\quad E^*(01 \parallel \text{first 59 bits of}\{E^*(110 \parallel X_{1i})\} \parallel \\
&\quad \text{first 59 bits of}\{E^*(111 \parallel X_{2i})\} \\
H(M) &= H_t
\end{aligned}
$$

In this scheme E^* is defined as Winternitz' construction and the strings 00, 01, 100, 101, 110 and 111 have been included to prevent the manipulation of weak keys.

2.4.9   N-hash Algorithm

N-hash is a hashing algorithm which produces a 128-bit hash value [Miyaguchi et al., 1989]. The algorithm, which was suggested by the designers of the FEAL block cipher algorithm, is based on a 128-bit encryption algorithm with the key length equal to the block length. The encryption algorithm is a Feistel type cipher with 16 rounds, and takes advantage of the f functions of FEAL. The N-hash algorithm uses yet another chaining scheme and is defined as follows in Figure 2.8.

$$
\begin{aligned}
H_0 &= IV \\
H_i &= E(M_i, H_{i-1}) \oplus H_{i-1} \oplus M_i \qquad i = 1, 2, \dots, t \\
H(M) &= H_t
\end{aligned}
$$

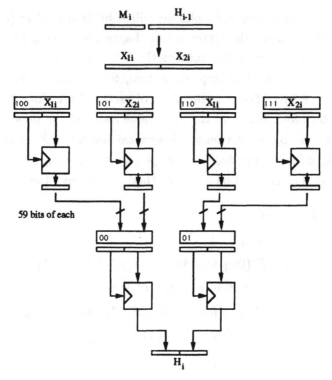

Figure 2.7: Merkle's Scheme

Biham and Shamir showed that the N-hash algorithm is susceptible to differential cryptanalysis and that it is possible to find collision messages for it [Biham and Shamir, 1991a].

2.4.10 MDC2 and MDC4

For modification detection in secure transactions, IBM proposed its MDC hashing scheme. There are two versions of this hashing scheme, namely,

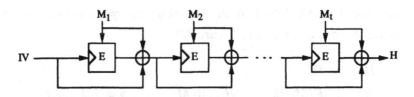

Figure 2.8: N-hash Structure Based on a Block Cipher Algorithm

MDC2 and MDC4. The former applies two DES encipherments per 8-byte input block, whilst the latter applies four DES encipherments. The MDC calculation procedure defines a one-way function based on Winternitz' proposal. Two different versions allow the user to make a trade-off between performance and security. As descriptions of the two proposals are beyond the scope of this overview, we give just a schematic presentation of MDC4, where better security is achieved at the expense of slower performance (see Figure 2.9). The idea is to first encrypt the message blocks using the previous hash result as the key, and then encrypt the hash block while the encrypted message serves as the key. In Figure 2.9, $M1_i$ and $M2_i$ are each 64 bits in length, and contain the left half and the right half of the 128-bit input message block, respectively. Note $H1_{i-1}$ and $H2_{i-1}$ form the key input of the 128-bit cryptographic procedure. The *mod1* function sets the bits 1 and 2 of input to 1 and 0, respectively, while the *mod2* function sets the bits 1 and 2 of its input to 0 and 1, respectively. These functions remove the symmetric structure of the above hash scheme, and also prevent manipulations based on a weakness of DES, that $\overline{E}(K, M) = E(\overline{K}, \overline{M})$. They also exclude weak keys of DES. As four DES encryptions are performed on each 128-bit message block, the scheme is less efficient than the previously mentioned ones. However, there has not yet been any attack on these schemes. Although the literature on the hashing schemes based on block ciphers is larger than we have presented here, the above schemes are representative of the types of proposals and the problems involved in them. There are collections of proposals based on block ciphers in [Akl, 1983] and [Meijer and Akl, 1982] for further reference.

2.5   Non-block-cipher-based Hash Functions

The other proposals for hash functions are those that do not take advantage of block cipher algorithms, but of functions that are complicated and difficult to invert. We call the second class of hash functions *non-block-cipher-based* hash algorithms. As the name implies, these hash functions include proposals based on one-way functions from number theory, e.g. the RSA, squaring, knapsack, or complicated software algorithms, or cellular automata schemes. In the following we describe some of the *non-block-cipher-based* hash schemes according to the kind of function employed.

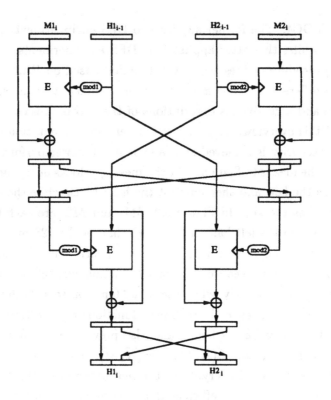

Figure 2.9: IBM's MDC4 Scheme

2.5.1   Cipher Block Chaining with RSA

For the first scheme, consider the RSA algorithm as the underlying one-way
function, and perform the Cipher Block Chaining mode of bock ciphers with
it. The description of the scheme is as follows:

$$
\begin{aligned}
H_0 &= IV \\
H_i &= (H_{i-1} \oplus M_i)^e \bmod N \qquad i = 1, 2, \ldots, t \\
H(M) &= H_t
\end{aligned}
$$

where N and e are public. A correcting block attack can compromise the
scheme by appending or inserting a correcting block to achieve a desired hash
value. A modified version of the above scheme is to add some additional

redundancy to the message to avoid a correcting block attack [Davies and Price, 1980]. To achieve a secure RSA, N should be at least 512 bits in length and as the result implementation of the above algorithm is very slow.

2.5.2   Schemes Based on Squaring

Davies and Price's Squaring Scheme

In order to speed up the above cipher block chaining algorithm with RSA, Davies and Price proposed the application of squaring instead of using the public exponent [Davies and Price, 1984]. Thus

$$H_i = (H_{i-1} \oplus M_i)^2 \bmod N$$

To avoid a correcting block attack, they suggested setting 64 bits of every message block to 0. However, Girault has shown that it is possible to find collision messages for the scheme [Girault, 1987]. To improve this scheme, the introduced redundancy should be of the order of 128 bits.

Girault also discussed several other schemes based on squaring. These schemes are listed as follows:

1. $H_i = H_{i-1} \oplus (M_i^2 \bmod N)$

   This scheme is vulnerable to attacks based on the permutation of blocks, insertion of an even number of blocks, insertion of zero blocks, or manipulations on small blocks where $M_i^2 < N$.

2. $H_i = M_i \oplus (H_{i-1}^2 \bmod N)$

   This method is vulnerable to attacks based on a correcting block.

3. $H_i = (H_{i-1} \oplus (M_i^2 \bmod N))^2 \bmod N$

   There is no gain in the execution time of this scheme, but on the other hand no attack on it has yet been given.

CCITT's Proposal

Appendix D of the X.509 recommendations of CCITT standards on secure message handling, proposes an algorithm for hashing based on squaring. The proposed scheme introduces 256 bits of redundancy to be distributed over every 256-bit message block by interleaving every four bits of the message with 1111, so that the total number of bits in each block becomes 512. Then the CBC mode of the exponentiation algorithm with exponent equal to 2 is run on the modified message. This scheme makes the four most significant bits of every byte in each block equal to 1. However, Coppersmith developed an attack to construct collision messages for this scheme when the hash scheme is used with an RSA signature scheme [Coppersmith, 1989].

Jueneman's Scheme

As squaring is the fastest possible exponentiation, Jueneman proposed several approaches based on squaring. His first proposal is similar to that of Davies and Price with the difference that the *exclusive-OR* operation is replaced by addition, and N is the prime number $2^{31} - 1$. This scheme results in a 32-bit hash value. However, the scheme has two weaknesses. First the hash value is rather short. Second, the scheme is vulnerable to the *meet-in-the-middle* attack. To obtain a 128-bit hash result, the designer proposed iterating the scheme four times [Jueneman et al., 1985]. This scheme is vulnerable to the generalized birthday attack, because of the common modulus of all the four iterations. The third version was to choose four different moduli N_j (for $j = 1$ to 4) equal to the four largest prime numbers smaller than $2^{31} - 1$ [Jueneman, 1987], [Jueneman, 1986]. The scheme can be described as follows. Divide the message into blocks of 128 bits length. Then split each message block M_i into four $M1_i$ to $M4_i$. A fifth block is constructed with selection of some bits from $M1_i$ to $M4_i$, as follows. $M5_i = (00 \parallel M1_{i_{31-26}} \parallel M2_{i_{31-24}} \parallel M3_{i_{31-24}} \parallel M4_{i_{31-24}})$ and the second bit of $M1_i$ to $M4_i$ is set to 0. For $j = 1$ to 4, the four functions $H_{j,i}$ are described as:

$$
\begin{aligned}
H_{j,i} = & \; [(H_{j\bmod 4, i-1} \oplus M1_i) - (H_{(j+1)\bmod 4, i-1} \oplus M2_i) + \\
& + (H_{(j+2)\bmod 4, i-1} \oplus M3_i) - (H_{(j+3)\bmod 4, i-1} \oplus M4_i) + \\
& + (-1)^{j+1} M5_i]^2 \bmod N_j
\end{aligned}
$$

However, it is reported in [Preneel et al., 1992] that Coppersmith has broken it in about 2^{32} operations with a correcting block attack that combines algebraic manipulations with a birthday attack.

Damgard's Squaring Scheme

Damgard, in his paper on the design principles for the construction of collision free hash functions [Damgard, 1989], described a scheme based on squaring to map a block of n bits into a block of m bits. The description of the scheme is as follows:

$$
\begin{aligned}
H_0 &= IV \\
H_i &= \text{extract m bits of } (00111111 \parallel H_{i-1} \parallel M_i)^2 \bmod N \\
H(M) &= H_t
\end{aligned}
$$

In the above scheme, *extract* is a function which extracts m bits from the result of the squaring function. To obtain a secure scheme m should be big enough to thwart the birthday attack. Moreover, the *extract* function should select bits for which finding colliding inputs is made difficult. One choice is to extract m uniformly distributed bits. However, for practical reasons, it is better to bind them together in bytes. Another possibility is to extract every fourth byte. In [Daemen *et al.*, 1991b], the authors reported that this scheme can be broken.

2.5.3 Schemes Based on Claw-Free Permutations

Damgard showed that it is possible to construct collision-free hash functions based on the existence of claw-free permutations [Damgard, 1987]. A claw-free set of permutations is a set of permutations $S = \{f_0, f_1, \ldots, f_{r-1}\}$, such that, for each x in the domain of f_i, it is easy to compute $f_i(x)$ for all $i = 0, \ldots, r-1$, but it is computationally intractable to create a *claw*, that is, to find a y such that, for some $i \neq j$, $f_i(x) = f_j(y)$. Damgard argued that the following scheme would yield a provable collision free-hash function.

$$
\begin{aligned}
H_0 &= IV \\
H_i &= f_{M_i}(H_{i-1}) \qquad i = 1, 2, \ldots, t \\
H(M) &= H_t
\end{aligned}
$$

Goldwasser, Micali and Rivest also showed that a similar structure with $r = 0$ would yield a secure signature scheme [Goldwasser *et al.*, 1988]. Damgard also proposed three further schemes, based on modular squaring, for the construction of claw-free permutations.

2.5.4 Schemes Based on the Knapsack Problem

As the *knapsack* problem is one of the problems in number theory which is considered to be difficult to invert, there have been some proposals made for it.

Damgard's Knapsack Scheme

Another scheme proposed by Damgard for the construction of collision-free hash functions is based on the application of the knapsack and can be described as follows. Choose at random numbers $a_1, \ldots, a_s$ in the interval $1, \ldots, N$, where s indicates the maximum length of a message to be expected in blocks. Damgard chooses $s = 256$ and $N = 2^{120} - 1$. Then the binary message $M_1, M_2, \ldots, M_s$ can be hashed as:

$$H(M) = \sum_{i=1}^{s} M_i a_i$$

This scheme would give an output in the length of 128 bits.

However, Camion and Patarin have shown that the above scheme is not secure [Camion and Patarin, 1991]. They demonstrate that a probabilistic algorithm with about 2^{32} computations can break the scheme; this number of operations is feasible with modern computer technology.

Impagliazzo and Naor's Scheme

Impagliazzo and Naor proposed a cryptographic subset sum function which can be applied for hashing schemes. The description of the scheme is as follows. Choose at random numbers $a_1, \ldots, a_n$ in the iterval $0, \ldots, N$, where n indicates the length of the message in bits, and $N = 2^\ell - 1$ where $\ell < n$. The binary message $M = M_1, M_2, \ldots, M_n$ corresponds to a subset $S \subset \{1, \ldots, n\}$ and can be hashed as:

$$H(M) = \sum_{i \in S} a_i \bmod 2^\ell$$

Impagliazzo and Naor have not mentioned any concrete values for the above scheme, but they have shown that it is theoretically sound.

2.5.5   Schemes Based on Cellular Automata

In a simple case, a cellular automaton consists of a line of cells or sites, each with value 0 or 1. These values are updated in a sequence of discrete time steps, according to a definite and fixed rule. Denoting the value of a cell at position i by a_i, a simple rule gives its new value as:

$$a_i' = \phi(a_{i-1}, a_i, a_{i+1})$$

where ϕ is a Boolean function which specifies the rule [Wolfram, 1986]. Despite the simplicity of their construction, many cellular automata schemes produce systems of considerable complexity.

Damgard's Scheme Based on Wolfram's Pseudorandom Bit Generator

Wolfram's pseudorandom bit generator consists of a one-dimensional cellular automaton of n bits. Let $x = x_0, x_1, \ldots, x_{n-1}$ be the input seed of the bit generator. The function $g(x_i)$ defines the value of i-th cell in the next time step. The function g is defined as follows:

$$g(x_i) = x_{i-1} \oplus (x_i \vee x_{i+1})$$

where $\vee$ stands for OR, and $\oplus$ means XOR. Denote the value of the i-th cell in the j-th time step by $g_j(x_i)$. The bit generator $b(x)$ starts from a random x and outputs the sequence $g_j(x_0)$. For $d > c$, let $b_{c-d}(x)$ denote the string

$$g_c(x_0), g_{c+1}(x_0), \ldots, g_d(x_0)$$

The hash function is defined as follows:

$$
\begin{aligned}
H_0 &= IV \\
H_i &= b_{c-d}(M_i \parallel H_{i-1} \parallel Z) \qquad i = 1, 2, \ldots, t \\
H(M) &= H_t
\end{aligned}
$$

where Z is a random value, added to make finding collision messages more difficult. As a concrete proposal, Damgard suggested $n = 512$, $r = 256$, $c = 257$ and $d = 384$. The proposed hash function will hash messages of arbitrary length into 128-bit strings. However Daemen, Govaerts, and Vandewalle showed how to cryptanalyze this scheme and find colliding messages [Daemen et al., 1991b].

Cellhash Scheme

In order to achieve a hash scheme that can be implemented on a chip, Deamen, Govaerts and Vandewalle [Daemen *et al.*, 1991b] proposed a hardware-oriented one-way hash function which is called *Cellhash*. The properties desired in the design of Cellhash were firstly to achieve a size of at least 128 bits to prevent a birthday attack; the result of cellhash is then 257 bits long. The second property was to design a function such that the diffusion of information could be guaranteed. The third was to achieve confusion of information so that the hash result depends on the bits of the message in a complicated way. The fourth property was to achieve a hash function which actually works at high speed. The computation of the cellhash is done as follows. First, 0's are appended to the message so that the length is at least 248 bits and congruent to 24 (mod 32). The number of bits added is represented in a byte subsequently appended. IV is the all-zero bit string of length 257. The computation of H_j from H_{j-1} is done under the key M_j, the j-th message block, and can be considered as a 5-step transformation. The calculations in each step are done simultaneously on all bits of H_{j-1}. Let $h_0, h_1, \ldots, h_{256}$ denote the bits of H_{j-1} and $m_0, m_1, \ldots, m_{255}$ denotes the bits of M_j.

$$
\begin{aligned}
&\text{Step1}: && h_i = h_i \oplus (h_{i+1} \vee \overline{h}_{i+2}) && 0 \leq i < 257 \\
&\text{Step2}: && h_0 = \overline{h}_0 \\
&\text{Step3}: && h_i = h_{i-3} \oplus h_i \oplus h_{i+3} && 0 \leq i < 257 \\
&\text{Step4}: && h_i = h_i \oplus m_{i-1}. && 1 \leq i < 257 \\
&\text{Step5}: && h_i = h_{10i} && 0 \leq i < 257
\end{aligned}
$$

There has not been any attack on this scheme yet.

2.5.6 Software Hash Schemes

There are numerous designs where the underlying one-way function is a block cipher, and some of them were shown in the previous section. As the purpose of a block cipher is different from that of a hash function, some dedicated software hash functions have been proposed to provide more efficient solutions.

MD4 and MD5

Rivest proposed MD4 for hashing [Rivest, 1990]. It is a software oriented scheme which is especially designed to be quite fast on 32-bit machines. The algorithm produces a 128-bit output; so it is not computationally feasible to produce two messages having the same hash value. The scheme provides diffusion and confusion of the input information, while it does not use any tables or *S*-boxes. MD4 has been placed in the public domain for review. The description of MD4 is beyond the scope of this overview.

The MD5 hashing algorithm is a strengthened version of MD4. It has more rounds and incorporates other revisions based on comments for the MD4 algorithm. There is not yet any known method of breaking MD4 or MD5. Den Boer and Bosselaers demonstrated an attack on the last two rounds of MD4 in [den Boer and Bosselaers, 1991]. Their work shows that, if the three-round MD4 algorithm is stripped of its first round, it is possible to find for a given input value two different messages hashing to the same output. There is also another work [Berson, 1992] which analyses any single round of MD5 separately.

HAVAL

HAVAL stands for a one-way hashing algorithm with variable length of output. It was designed at the University of Wollongong by Zheng, Pieprzyk, and Seberry [Zheng *et al.*, 1992]. It compresses a message of an arbitrary length into a digest of 128, 160, 192, 224 or 256 bits. The security level can be adjusted by selecting 3, 4, or 5 passes. The HAVAL structure is based on MD4 and MD5. Unlike MD4 and MD5 whose basic operations are being done using functions of three Boolean variables, HAVAL employs five Boolean functions of seven variables (each function serves a single pass). All functions used in HAVAL are highly nonlinear, 0-1 balanced, linearly inequivalent, mutually output-uncorrelated and satisfy the Strict Avalanche Criterion (SAC).

The structure of HAVAL is much more complex than MD5 and the authors argue that HAVAL with five passes, is more secure than MD5. The experiments showed that HAVAL with 3 passes is 60% faster than MD5, it is 15% faster than MD5 when HAVAL applies 4 passes and it is as fast as MD5 when it has 5 passes. The authors mentioned an even faster HAVAL version which is based on Boolean functions of five variables [Charnes and Pieprzyk, 1992]. There has been no attack on the scheme to date.

Snefru

Another software hashing scheme is *Snefru*, proposed in [Merkle, 1990b]. Merkle suggested this scheme as a hashing scheme which is easy to implement, resistant to cryptographic attacks, and is fast when implemented in software. Snefru produces a hash result of 128 bits and takes advantage of 8 S-boxes. The original scheme of Snefru had two rounds. However, Biham and Shamir have shown how to create an unlimited number of pairs hashing to the same 128-bit hash value with a two-round *Snefru* [Biham and Shamir, 1991b]. This resulted in a later proposal for Snefru which consisted of four rounds. Although Snefru provides greater flexibility in selecting input and output block sizes, MD4 was already slightly faster than Snefru with two passes. The change to four passes means MD4 is now over twice as fast as Snefru. Altogether, if MD4 proves secure, it is more attractive as a standard for hashing since it has a better performance [Merkle, 1990a].

2.5.7 Matrix Hashing

Matrix algebra is another area which has been used in the construction of hashing algorithms. One such proposal is called the Random Matrix Hashing Algorithm [Banieqbal and Hilditch, 1990]. The algorithm considers its input to be a $1 \times m$ row vector of bits, and its output to be an $n \times 1$ column vector of bits. The algorithm consists of choosing a fixed $m \times n$ random binary matrix, and multiplying it by the input vector. The matrix should be kept secret as a key. Moreover, it can be chosen in such a way that the scheme is invertible if required. The algorithm can be sized to any bit length for input or output or both.

Another scheme, invented by Harari also uses as key a random $t \times t$ matrix A, with t the number of m-bit blocks of the plaintext M [Harari,

1984]. The hash value is computed as

$$H(M) = M^T A M = \sum_{i<j} a_{ij}.x_i.x_j$$

The scheme, however, is insecure under a chosen message attack.

2.5.8   Schnorr's FFT Hashing Scheme

Schnorr proposed an efficient algorithm that hashes messages of arbitrary bit length into a 128-bit hash value [Schnorr, 1991]. The algorithm consists of two stages, a discrete Fourier transformation and a polynomial recursion over a finite field. The message is padded so that its length in bits becomes a multiple of 128. It is recommended that the message is appended so that a single 1 bit is followed by a suitable number of 0 bits which are also followed by a binary representation of the message length in bits. Let the padded message consist of n blocks $M_1, \ldots, M_n$ each of which is 128 bits long. The algorithm for the hash function h is:

$$\begin{aligned} H_0 &= IV = 0123456789ABCDEFFEDCBA987654321 \\ H_i &= g(M_i \parallel H_{i-1}) \\ H(M) &= H_t \end{aligned}$$

g is a hash algorithm with input size of 256 bits and output size of 128 bits. g uses the discrete Fourier transform FT_8. It is given that $FT_8(a_0, \ldots, a_7) = (b_0, \ldots, b_7)$ with

$$b_i = \sum_{j=0}^{7} 2^{4ij} a_j \pmod{p} \qquad \text{for } i = 0, \ldots, 7$$

Let p be the prime $p = 65537 = 2^{16} + 1$. Let the input to g be denoted by $(e_0, \ldots, e_{15}) \in \{0,1\}^{256}$; then the description of g is as follows.

1. $(e_0, e_2, e_4, \ldots, e_{14}) = FT_8(e_0, e_2, e_4, \ldots, e_{14})$

2. For $i = 0, \ldots, 15$ do

   $$e_i = e_i + e_{i-1}e_{i-2} + e_{e_{i-3}} + 2^i \pmod{p}$$

   (The indices $i, i-1, i-2, i-3, e_{i-3}$ are taken modulo 16)

3. Repeat steps 1 and 2

However, Daemen, Bosselaers, Govaerts, and Vandewalle [Daemen *et al.*, 1991a], and also Baritaud and Gilbert [Baritaud and Gilbert, 1992] presented their attacks on the scheme. They showed that it is possible to construct collisions for Schnorr's FFT Hashing scheme. This resulted in a later proposal based on FFT, where the weaknesses discovered are removed. The improved version is called FFT-Hash II and is detailed in [Schnorr, 1992]. At the Crypto'92 Conference, Serg Vandery showed that this version is not secure either [Vandery, 1992].

2.6 Design Principles for Hash Functions

As we mentioned earlier, a hash function is called *collision free* if it maps messages of any length to strings of some fixed length, such that finding x, y with $h(x) = h(y)$ is a hard problem. Many of the difficulties in giving proofs for known constructions, arise from the fact that things seem to get more complex as the lengths of the messages hashed increase. On the other hand, a hash function is of no use if we are not allowed to hash messages of arbitrary lengths. Damgard presents two methods in [Damgard, 1989] to remove this difficulty. He shows that the ability to cut just one bit off the length of a message in a collision-free way implies an ability to hash messages of an arbitrary length. The methods are basic design principles which can be used as guide for designing hash functions. One approach allows hashing of long messages to be implemented serially, while the other approach allows parallel hashing of long messages. The serial method is the same as Merkle's *meta method*, which was invented independently and has been presented in [Merkle, 1989b].

2.6.1 Serial Method

Let f be a fixed-size, collision-free, hash function mapping ℓ bits to m bits. Then a collision-free hash function H, which maps strings of arbitrary length to m bit strings can be constructed as follows. The input $M \in \{0,1\}^*$ is split in blocks of the size $\ell - m - 1$ bits. If the block is incomplete, it is padded with 0's. Let d be the number of 0's needed. The binary representation of d, prefixed with an appropriate number of 0's, would be appended as an extra block. Assume the length of the text after padding is n. Generate a sequence of m bit blocks $h_0, h_1, \ldots, h_{\frac{n}{\ell-m}+1}$, by:

$$
\begin{aligned}
h_1 &= f(0^{m+1} \parallel M_1) \\
h_{i+1} &= f(h_i \parallel 1 \parallel M_{i+1}) \qquad i = 1, 2, \ldots, \frac{n}{\ell - m} \\
H(M) &= h_{\frac{n}{\ell-m}+1}
\end{aligned}
$$

2.6.2   Parallel Method

This method would allow parallel computation of the hash value on several processors. If, for example, c processors co-operate, they would achieve a speed increased by a factor of c. In some references such as [Preneel et $al.,$], this method is called the $tree$ approach to hashing functions.

Let f be a fixed-size, collision-free, hash function mapping m bits to t bits. Then a collision free hash function H which is implemented in a parallel way and maps strings of arbitrary length to t bit strings can be constructed as follows. Let a message M of length n be given. The message is padded with a number of 0's so that the resulting bit string has length equal to $2^k m$ for some k.

$$
\begin{aligned}
h_i^1 &= f(M_{2i-1} \parallel M_{2i}) & i &= 1, \ldots, 2^{k-1} \\
h_i^j &= f(h_{2i-1}^{j-1} \parallel h_{2i}^{j-1}) & i &= 1, \ldots, 2^{k-j} \qquad (j = 2, \ldots, k-1) \\
H(M) &= f(h_1^{k-1} \parallel h_2^{k-1})
\end{aligned}
$$

The final hash value is $H(M)$ for the message M.

2.7   Conclusions

In this chapter we reviewed the basic definitions of hash functions. We presented several classifications for hash functions. The first classification divides the proposed hashed functions according to their level of security. In this classification they are divided into weak and strong one-way hash functions. The second classification was more concerned with the technical issue of whether a private key was involved in the scheme or not. The third classification was concerned with the structure of the hashing algorithm itself, and it considered whether a block cipher had been applied as the underlying one-way function. We reviewed various proposals, and divided them according to the third classification.

Although this chapter does not cover all the proposals, whether based on a block cipher or on other one-way functions, it gives a representative overview of the type of proposals and the problems associated with them. At the end, we remark that our aim is to develop some design rules for the construction of hash functions where they are considered as *block-cipher-based* and *non block-cipher-based*.

Chapter 3

Methods of Attack on Hash Functions

3.1 Introduction

The best method to evaluate a hash scheme is to see what attacks an adversary may perform to find two messages that map to the same hash value. The hashing algorithm produces, as the hash value, a fixed length 'random' number which depends on all the bits of the message. In general, it is assumed that the adversary knows the hash algorithm. As a conservative approach, it is assumed that he or she can perform an adaptive chosen message attack, where he or she may choose messages, ask for their hash values, and try to compute messages with the same hash value. There are several methods for using such pairs in order to attack a hash scheme and to calculate colliding messages. Some methods are general and can be applied against any hash scheme. The so-called *birthday attack* is such a method and can be applied against any type of hash scheme. Other methods are applicable against only special groups of hash schemes. Some of these special attacks can be launched against a wide range of hash functions. For example, the so-called *meet-in-the-middle attack* can be launched against any scheme that uses some sort of block chaining in its structure. Others can be launched only against smaller groups. For example, the so-called *correcting block attack* is applied mainly against hash functions based on modular arithmetic.

Furthermore, some hash schemes have been broken with methods which are only applicable to those particular hash schemes. Such attacks are not included in this chapter; however, it was mentioned in Chapter 2 how a hash scheme can be broken in a special way.

In this chapter, we give a brief explanation of these *general attacks* and *special attacks*.

3.2   General Attacks

In Subsection 2.4.1 we introduced Rabin's hashing scheme. The scheme is an efficient hash function based on a block cipher. Rabin used DES as the block cipher. As DES transforms 64-bit plaintext blocks to 64-bit ciphertext blocks, the proposed scheme provides a 64-bit hash value. Later, Yuval showed in [Yuval, 1979] that this scheme is subject to the so-called *birthday attack*. The idea behind the attack originates from a famous problem from *probability theory*, called the *birthday* problem. The paradox can be stated as follows: What is the minimum number of pupils in a classroom so that the probability that at least two pupils in this classroom have the same birthday is greater than 0.5? The answer to this question is 23, which is much smaller than the value one might suggest by intuition. The justification for this result is as follows. Suppose that the pupils are entering the class one at a time. The probability that the birthday of the first pupil is a specific day of the year is equal to $\frac{1}{365}$. The probability that the birthday of the second pupil is not the same as the first one is equal to $1 - \frac{1}{365}$. If the birthdays of the first two pupils are different, the probability that the birthday of the third pupil is different from the first one and the second one is equal to $1 - \frac{2}{365}$. Consequently, the probability that t students have different birthdays is equal to $(1 - \frac{1}{365})(1 - \frac{2}{365}) \ldots (1 - \frac{t-1}{365})$, and the probability that at least two of them have the same birthday is

$$P = 1 - (1 - \frac{1}{365})(1 - \frac{2}{365}) \ldots (1 - \frac{t-1}{365})$$

It can be easily computed that for $t \geq 23$, this probability is bigger than 0.5.

The idea of the above problem can be employed for attacking hash functions. Suppose that the number of bits of the hash value is n. An adversary generates r_1 variations on a bogus message and r_2 variations on a genuine message. The probability of finding a bogus message and a genuine message that hash to the same result can be approximated by

$$P \approx 1 - e^{-\frac{r_1 r_2}{2^n}}$$

where $r_2 \gg 1$ [Ohta and Koyama, 1990]. When $r_1 = r_2 = 2^{\frac{n}{2}}$, the above probability is about 0.63. Jueneman has shown in [Jueneman, 1986] that for $n = 64$ the processing and sorting requirements are feasible in reasonable time with today's computing resources. On the other hand, a memory-time trade-off is also possible. It is usually recommended that the hash value should be around 128 bits to achieve security against a *birthday attack*.

This method of attack does not take advantage of the structural properties of the hash scheme or its algebraic weaknesses. In other words, it can be launched against any hash scheme. In addition, it is assumed that the hash scheme assigns to a message a value which is chosen with a uniform probability among all the possible hash values. Note that if there is any weakness in the structure or certain algebraic properties of the hash scheme, or the hash values do not have a uniform probability distribution, then generally it would be possible to find colliding messages with a better probability and fewer message-hash value pairs.

The *birthday attack* is a general method against authentication schemes, even if the hash function is applied to encrypted data or is evaluated under the control of a private key. Ohta and Koyama explain how the *meet-in-the-middle attack*, which is a version of the birthday attack, can be employed against signature schemes, where a signatory can forge a bogus message for his own signature, or an adversary can offer the signer a message he or she is willing to sign and replace it later with a bogus message [Ohta and Koyama, 1990].

3.3   Special Attacks

Unlike the birthday attack, which can be launched against any hashing scheme, there are some methods of attack that can be launched against only some groups of hash functions. We review these methods in this section.

3.3.1   Meet-in-the-middle Attack

Meet-in-the-middle attack is a variation of the birthday attack, but instead of comparing the hash values, the intermediate variables in the chaining are compared. The attack can be launched against schemes which employ some sort of block chaining in their structure. In contrast to birthday attack, meet-in-the-middle attack enables an attacker to construct a bogus message with a desired hash value. In this attack the message is divided into two parts. The attacker generates r_1 variations on the first part of a bogus message. He starts from the initial value and goes forward to the intermediate stage. He also generates r_2 variations on the second part of the bogus message. He starts from the hash result and goes backward to the intermediate stage. The probability of a match in the intermediate stage is the same as the probability of success in the birthday attack.

Nishimura and Sibuya described three variations of this attack in [Nishimura and Sibuya, 1990]. They argued that crediting the high probability of success in the meet-in-the-middle attack to the classical birthday problem is not exact and is misleading. However, they conceded that the asymptotic conclusions in the literature are correct. They considered three matching models and called them the model A, B and C. Later they calculated the exact probabilities of success for each attack with the specified matching model. They based their discussion on the assumption that the encryption and decryption functions to be used were random. As in many hashing schemes, DES was used as the underlying cryptosystem. The question whether DES has some algebraic structure or can be considered random, has been studied extensively in [Hellman *et al.*, 1976]. We give a detailed discussion of which cryptosystems can be considered random in Chapter 4.

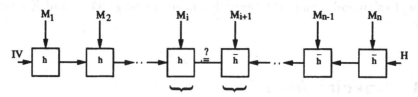

Figure 3.1: Meet-in-the-middle Attack, Model B

Model A is a typical attack and was described previously. In model B most of the bogus message is kept fixed, but r_1 variants of the (bogus) message block, one stage before the intermediate stage, and r_2 variants of the bogus message block, after the intermediate stage are provided. Then the forward and backward procedures are applied. Figure 3.1 shows the intermediate stage and message blocks one stage before and after it.

Model B attack is also an effective attack against the block cryptosystem itself [Nishimura and Sibuya, 1990]. The block ciphers most resistant to model B attacks are those which are secure against chosen plaintext/ciphertext attacks[1].

In model C attack, the forward sequences are generated as in the model A attack, whilst the backward sequences are generated as in the model B attack. So r_1 variants of the first part of the bogus message are provided, and r_2 variants of the bogus message block, after the intermediate stage are made. Nishimura and Sibuya suggested that model C attack is effective against hashing schemes based on a cipher block chaining mode of the block cipher.

The meet-in-the-middle attack can be prevented by avoiding schemes which are invertible. Winternitz' scheme is an effort to get around this attack [Winternitz, 1983].

3.3.2   Generalized Meet-in-the-middle Attack

To avoid the meet-in-the-middle attack, some authors [Davies and Price, 1980] [Davies and Price, 1984] considered two-fold iterated schemes. These include iteration of a hashing scheme with two different initial values, and/or repeating the message twice and then applying the hash scheme. However, Coppersmith [Coppersmith, 1985] and Girault et al. [Girault *et al.*, 1988] extended the meet-in-the-middle attack to break not only the two-fold but also p-fold iterated schemes. They called their method the *generalized meet-in-the-middle* attack and showed that it requires only $O(10^p.2^{\frac{n}{2}})$ operations (n is the number of bits in the hash string). In their method of attack, a trade-off is made between time and storage.

---

[1]A more detailed explanation of the chosen plaintext/ciphertext attack and also of the chosen plaintext attack will be given in Chapter 4.

3.3.3 Correcting Block Attack

In this attack, the bogus message is concatenated with a block in order that the hash result is corrected and attains the desired value. This attack is often applied to the last block and is called *correcting last block* attack, although it can be applied to other blocks as well. In [Mitchell, 1989] and [Mitchell and Walker, 1988], such an attack against a hash scheme based on the CBC mode of DES has been described. Hash functions based on modular arithmetic are especially sensitive to the correcting last block attack [Preneel *et al.*, 1992]. The introduction of redundancy into the message in these schemes, makes finding a correcting block with the necessary redundancy difficult, although, it makes the scheme less efficient. We should mention here that the difficulty of finding a correcting block depends on the nature of the introduced redundancy. For example, Coppersmith has shown in [Coppersmith, 1989] that the redundancy proposed by the CCITT, for the modular squaring hashing scheme, does not provide a secure hash scheme.

3.3.4 Attacks Depending on Algorithm Weaknesses

As we mentioned in Section 2.4.3, a hashing scheme based on a block cipher algorithm in cipher block chaining or cipher feedback or output feedback mode of operation can be compromised by insertion, permutation and substitution of the blocks. These attacks take advantage of the algebraic structure of the hashing scheme. Miyaguchi, Ohta, and Iwata showed in [Miyaguchi *et al.*, 1990] how to compromise many hash schemes by using the algebraic properties of the structure of each hashing scheme and certain weaknesses of the underlying block cipher. For example, some well known weaknesses of DES which have been exploited are as follows:

1. DES is symmetric under complementation, that is,

$$C = DES(K, M) \rightarrow \overline{C} = DES(\overline{K}, \overline{M})$$

   This weakness allows the construction of trivial collisions.

2. DES has weak and semi-weak keys. There are 4 weak keys, for which encryption equals decryption, that is $DES(K, M) = DES^{-1}(K, M)$.

There exist also 6 pairs of semi-weak keys, for which

$$DES(K_2, DES(K_1, M)) = M \ .$$

3. DES has key collisions. A collision is a pair of keys K_1, K_2 such that $DES(K_1, M) = DES(K_2, M)$ for a message.

3.3.5   Differential Cryptanalysis

Eli Biham and Adi Shamir have developed a method for attacking block ciphers, which they call *differential cryptanalysis* [Biham and Shamir, 1990]. This attack is a general method for attacking cryptographic algorithms. It has exposed the weaknesses in many cryptographic algorithms, including Snefru. Snefru is a software hash function proposed by Merkle [Merkle, 1990c], [Biham and Shamir, 1991b]. Recently, it has also been applied successfully to break one round of the MD5 hash scheme by Berson [Berson, 1992]. The differential cryptanalysis attack takes advantage of the non-uniform probability distribution of the output caused by non-random S-boxes. A description of the attack is beyond the scope of this book; however the interested reader is referred to [Biham and Shamir, 1990], [Biham and Shamir, 1991a] and [Biham and Shamir, 1991b] for further information.

3.4   Conclusions

In this chapter, we reviewed some methods of attack on hashing algorithms. Differential cryptanalysis, the correcting block attack, and attacks depending on an algorithm's weak points are based on non-random behaviour of the hash scheme. On the other hand, the birthday attack and the meet-in-the-middle attack assume the hashing scheme is random, and they try to exploit the small bit length of the hash value (see [Nishimura and Sibuya, 1990]). For a hash scheme to be considered random in a birthday attack, it is enough for the scheme to be secure against chosen message attacks. In the meet-in-the-middle attack, an attacker starts from some initial value. Having a message of several blocks, he or she performs the hash scheme on the initial value and the first block and goes forward to reach a middle point. Then he or she starts from the final value, i.e., the hash value, and goes backwards to reach the middle point. He provides many variations of the message and repeats the above procedure on each. If the middle-stage values of two of the messages match, then two 'colliding' messages are found and the hash scheme is successfully attacked. As we will explain in the next chapter, this type of attack is a version of chosen plaintext/ciphertext attack on the underlying block cipher, where a cryptanalyst is allowed not only to choose plaintexts of his own choice and see the corresponding ciphertext, but also to choose ciphertext of his own choice and see the corresponding plaintext. For a hash scheme to be considered random in the meet-in-the-middle attack, the underlying block cipher should behave like a random permutation against chosen plaintext/ciphertext attack. In Chapters 4, 5, and 6, design rules for the development of a block cipher which is secure against chosen plaintext/ciphertext attack will be discussed.

Chapter 4

Pseudorandomness

4.1 Introduction

Block ciphers have been used as the underlying one-way function in the construction of hash functions, because of their ease of implementation. Some designers of hash algorithms have even proposed constructing $2n$-bit hash functions from n-bit block ciphers. However, Lai and Massey suggested that for a block-cipher-based hash scheme any attack on the block cipher itself implies an attack of the same type on the hash scheme with the same computational complexity [Lai and Massey, 1992]. Hence, block-cipher-based hash schemes may be vulnerable to attacks based on the exploitation of the algebraic properties of the underlying block cipher. Furthermore, if the block length of the underlying block cipher is rather short or it does not behave like a random transformation, then the hash scheme is vulnerable to attacks of the same type with the same computational complexity.

The meet-in-the-middle attack can be considered as a version of chosen plaintext/ciphertext attack against the block cipher, where a cryptanalyst is allowed not only to choose plaintext of his own choice and see the corresponding ciphertext, but also to choose ciphertext of his own choice and see the corresponding plaintext. If the block cipher behaves like a random permutation against chosen plaintext/ciphertext attack, then a hash scheme based on it, is secure against the meet-in-the-middle attack. Unfortunately, the known block ciphers are only claimed to be secure against chosen plaintext attacks, and none of them claim to be secure against chosen plaintext/ciphertext attack.

The design of most known block ciphers is based on the theoretical work of Shannon [Shannon, 1949b], [Shannon, 1949a]. He suggested that consecutive rounds of confusion and diffusion would provide a strong cryptographic algorithm. DES and most of the known block ciphers take advantage of Feistel type permutations. The definition of Feistel type permutations will be given later in Section 4.6. Such a permutation involves a function controlled by a key to provide the desired confusion and diffusion. On the other hand, it should be mentioned that the design rules for DES were never published. Luby and Rackoff showed that three rounds of Feistel type permutations, with three different random functions, would yield a block cipher which can be shown to be secure against chosen plaintext attack [Luby and Rackoff, 1988]. Although the functions employed in DES, i.e. the S boxes, are by no means random functions, Luby and Rackoff considered their result to be a justification for the application of a Feistel type permutation in the design of DES. In other words, although they did not examine the S boxes of DES, they showed that the structure applied in the design of DES is a sound structure for the design of block ciphers which are secure against chosen plaintext attack. In a similar vein, in Chapters 4, 5, and 6, we develop a structure which we show is secure against chosen plaintext/ciphertext attack. This chapter is devoted to preliminary definitions. We define what is meant by pseudorandomness, and when a generator can be distinguished from a truly random one. Then, the definitions for pseudorandom bit generators, pseudorandom function generators, and pseudorandom permutation generators are given. These definitions are based on a complexity-theoretic approach. We use circuits to evaluate whether a permutation generator is pseudorandom. These circuits model the chosen plaintext attack. As we wish to develop structures secure against the chosen plaintext/ciphertext attack, we also describe a circuit model for this attack. Permutation generators that are secure in this model are called super-pseudorandom.

4.2   Notation

The notations we use are similar to these in [Pieprzyk, 1991]. The set of all integers is denoted by N. Let $\Sigma = \{0,1\}$ be the alphabet we consider. For $n \in N$, Σ^n is the set of all 2^n binary strings of length n. The concatenation of two binary strings x, y is denoted by $x \parallel y$. The bit by bit exclusive-OR of x and y is denoted by $x \oplus y$. By $x \in_r S$, we mean that x is chosen from a set S uniformly at random.

4.3   Indistinguishability

Classical pseudorandom generators are deterministic algorithms with well defined mathematical structures that output numbers or binary strings that look like random ones. Statistical tests provide us with a useful tool for testing the quality of pseudorandom generators.

Consider two different message sources S and S' with their respective probability distributions p and p' over Σ, where Σ is the set of elementary messages. Given that one source is truly random and the other is not, an observer, having access to the outputs of the sources, tries to distinguish between them. Figure 4.1 depicts this scenario.

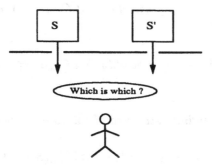

Figure 4.1: Distinguishing Two Sources S and S'

If the observer can gather enough occurrences of the elementary messages Σ, it may be possible for him to distinguish its source. Such message sources are called *classical* [Pieprzyk, 1990]. When the number of elementary messages is so large that it would be impossible to collect enough information about any elementary message, the previous approach fails. In this case, it is assumed that the observer is able to collect a polynomial number of elementary messages from the sources, or that the observer has polynomially limited computing power and polynomially limited time. Having such a resource, he uses an algorithm to process the collected information and to give the final decision as a binary output. When the algorithm decides that the source is a truly random one, it outputs **0**, or outputs **1** otherwise. Such an algorithm is called *a distinguisher*. It is noteworthy that the distinguisher can give its decision only with some probability. The more samples it gathers from the sources, the more probable that the decision is correct. However, it should be emphasized that the distinguisher can only access a polynomial number of samples.

Yao in his seminal paper [Yao, 1982] formally defined a distinguisher as follows:

Definition 4.1 *Let S, S' be two sources. A distinguisher C_n is a probabilistic polynomial time algorithm with the following properties.*

- *For any input (n, α), where $\alpha = (x_1, x_2, \ldots, x_{n^k})$ is a sequence of n^k outputs of S, the algorithm C_n halts in time $O(n^t)$ and gives a Boolean output $C_n(\alpha)$.*

- *$\mathrm{Prob}[C_n(S) = 1]$ is the probability that $C_n(\alpha) = 1$ when α is generated by S.*

- *$\mathrm{Prob}[C_n(S') = 1]$ is the probability that $C_n(\alpha) = 1$ when α is generated by S'.*

- *There exists an infinite sequence of values $n_1 < n_2 < \ldots$ such that*

$$| \mathrm{Prob}[C_n(S) = 1] - \mathrm{Prob}[C_n(S') = 1] | > \epsilon$$

for some fixed t, k and for any $\epsilon > 0$ where $n = n_1, n_2, \ldots$.

The equivalent of the above distinguisher can also be defined in terms of probabilistic Boolean circuits. As Luby and Rackoff used such circuits to define distinguishers, we prefer to use this model to agree with their approach. The Boolean circuit equivalent to the distinguisher algorithm can be described as follows:

Definition 4.2 *Let S, S' be two sources. A distinguisher C_n for (S, S') is an acyclic probabilistic circuit which contains Boolean gates, i.e., AND, OR and NOT gates, constant gates, i.e., '0' and '1', and accepts n^k n-bit inputs from the message source and randomly selected inputs such that the following conditions hold.*

- *For any input (n, α), where $\alpha = (x_1, x_2, \ldots, x_{n^k})$ is a sequence of n^k outputs of S, the circuit C_n gives a Boolean output $C_n(\alpha)$.*

- *The size of the circuit is less than or equal to n^t and is measured by the total number of connections inside the circuit.*

- *$\mathrm{Prob}[C_n(S) = 1]$ is the probability that $C_n(\alpha) = 1$ when α is generated by S.*

- *$\mathrm{Prob}[C_n(S') = 1]$ is the probability that $C_n(\alpha) = 1$ when α is generated by S'.*

- *There exists an infinite sequence of values $n_1 < n_2 < \ldots$ such that*

$$| \mathrm{Prob}[C_n(S) = 1] - \mathrm{Prob}[C_n(S') = 1] | > \epsilon$$

for some fixed t, k and for any $\epsilon > 0$ where $n = n_1, n_2, \ldots$.

The definition for indistinguishability can be given as follows, using the above distinguishing circuits for evaluation.

Definition 4.3 *Two sources S and S' are said to be indistinguishable if there exists no distinguisher for them.*

4.4   Pseudorandom Bit Generators

A bit generator is a deterministic algorithm which extends an n-bit input, known as a seed, to a bigger string of $O(n^k)$ bits. The definition for bit generators can be given formally as follows:

Definition 4.4 *Let l be a polynomial with $l(n) > n$. A bit generator is a deterministic polynomial-time function g that upon receiving an n-bit input as a seed, runs in polynomial time and extends the seed into a sequence of $l(n)$ bits $b_1, b_2, \ldots, b_{l(n)}$ as the output.*

A bit generator is called pseudorandom if, upon receiving a random n-bit seed for sufficiently large n, the corresponding generator is indistinguishable from a truly random one. The definition is given formally as follows:

Definition 4.5 *A bit generator g_n is pseudorandom if for large enough n and for any distinguisher C_n,*

$$| \operatorname{Prob}[C_n(g_n) = 1] - \operatorname{Prob}[C_n(R) = 1] | \leq \frac{Q(n)}{2^n}$$

where $\operatorname{Prob}[C_n(g_n) = 1]$ is the probability that the distinguisher C_n outputs 1, if an n-bit string is selected randomly and uniformly from all n bit strings as the seed to the bit generator and the distinguisher examines the n^k-bit string of the bit generator, and $\operatorname{Prob}[C_n(R) = 1]$ is the probability that the distinguisher outputs 1 if the n^k-bit string is selected randomly and uniformly from all possible strings, and $Q(n)$ is any polynomial in n.

Informally, g_n is pseudorandom if there is no polynomial (in n) size circuit, or no polynomial time algorithm which can significantly distinguish the $l(n)$-bit string of the output of the bit generator from a string randomly chosen from the set of all $l(n)$ bit strings, for infinitely many n.

Since any distinguisher is a specific test, the above definition can be stated that, *a bit generator is pseudorandom if it passes all polynomial-time tests, for large enough n.* Yao proved that a bit generator passes any polynomial size test if the output bits are unpredictable or the output string passes the *next bit* test. If given the generator g and the first s output bits of the bit generator $b_1, \ldots, b_s$ (note that the input seed is kept secret), it is not feasible computationally to predict the $(s+1)$th bit of the output string, it is said that the generator passes the *next bit* test. The following theorem is derived from [Yao, 1982]. It has been stated by Blum, Micali, Alexi, Chor, Goldreich, Schnorr, and Goldwasser in a different form ([Blum and Micali, 1984],[Alexi et al., 1988],[Goldreich et al., 1986]).

Theorem 4.1 *Let g be a polynomial bit generator, then the following statements are equivalent:*

- *g passes the next bit test.*

- *g is indistinguishable from a truly random bit generator.*

In other words, the indistinguishability test is equivalent to the unpredictability test.

All practical implementations of pseudorandom bit generators are based on functions which are conjectured to be one-way, where a one-way function, informally speaking, is a function which is easy to compute but hard to invert [Goldreich and Levin, 1989]. A formal definition of one-way functions will be given in Chapter 7. Unfortunately, complexity theory has not yet provided the answer to the fundamental question as to whether one-way functions exist. The relation between pseudorandom bit generators and is given in the following theorem of [Levin, 1987].

Theorem 4.2 *There exists a pseudorandom bit generator if there exists a one-way function.*

4.5   Pseudorandom Function Generators

In this section we present the notion of pseudorandom function generators. By a function f, we mean a transformation from Σ^n to Σ^n. The set of all functions on Σ^n is denoted by H_n, that is, $H_n = \{f \mid f : \Sigma^n \to \Sigma^n\}$, and it consists of 2^{n2^n} elements. The composition of two functions f and g is defined as $(f \circ g)(x) = f(g(x))$. The i-fold composition of f is denoted by f^i. A function f is a permutation if it is a one-to-one and onto function. The set of all permutations on Σ^n is denoted by P_n and it consists of $2^n!$ elements.

A function generator is a collection of functions with two properties: indexing and polynomial time evaluation. The precise definition of function generators is given below.

Definition 4.6 *Let $l(n)$ be a polynomial in n, a function generator $F = \{F_n : n \in N\}$ is a collection of functions with the following properties:*

- *Indexing: Each F_n specifies for each k of length $l(n)$ a function $f_{n,k} \in H_n$.*

- *Polynomial-time evaluation: Given a key $k \in \Sigma^{l(n)}$, and a string $x \in \Sigma^n$, $f_{n,k}(x)$ can be computed in polynomial time in n.*

A pseudorandom function generator is a function generator that cannot be distinguished from a truly random one. In other words, it is a collection of functions on n-bit strings that cannot be distinguished from the set of all functions on n-bit strings. To determine whether a collection of functions can be distinguished from the set of all functions, distinguishing circuits for functions are used, which are similar to distinguishing circuits for bit generators but are more powerful. They are, in fact, oracle circuits. The exact definitions of oracle circuits and distinguishing circuits for functions and pseudorandom function generators are given below.

Definition 4.7 *An oracle circuit C_n is an acyclic circuit which contains Boolean gates of the type AND, OR and NOT, and constant gates of the type zero and one, and a particular kind of gates named oracle gates. Each oracle gate has an n-bit input and an n-bit output and is evaluated using some function from H_n. The oracle circuit C_n has a single bit output.*

Definition 4.8 *The size of an oracle circuit C_n is the total number of connections between gates, Boolean gates, constant gates and oracle gates.*

Definition 4.9 *A distinguishing circuit family for a function generator F is an infinite family of circuits $\{C_{n_1}, C_{n_2}, \ldots\}$, where $n_1 < n_2 < \ldots$, such that for some pair of constants c_1 and c_2 and for each $n \in \{n_1, n_2, \ldots\}$ there is a circuit C_n with the following properties.*

- *The size of C_n is less than or equal to n^{c_1}.*

- *If $\mathrm{Prob}\{C_n[H_n] = 1\}$ is the probability that the output bit of C_n is one when a function is randomly selected from H_n and used to evaluate the oracle gates and if $\mathrm{Prob}\{C_n[F_n] = 1\}$ is the probability that the output bit of C_n is one when a key k of length $l(n)$ is randomly chosen and $f_{n,k}$*

*is used to evaluate the oracle gates, then the distinguishing probability
for C_n is greater than or equal to $\frac{1}{n^{c_2}}$, that is,*

$$| \operatorname{Prob}\{C_n[H_n] = 1\} - \operatorname{Prob}\{C_n[F_n] = 1\} | \geq \frac{1}{n^{c_2}}$$

Definition 4.10 *A function generator F is pseudorandom if there is no distinguishing circuit family for F.*

In other words, a distinguishing circuit for a function generator can be described as an algorithm that gathers a polynomial number of the inputs, for inputs of its own choice, to oracle gates with the function f. If the distinguisher decides that the f has been selected from F_n, it outputs **1**. If it decides otherwise, it outputs **0**, meaning that it has decided that f has been randomly selected from H_n. If the probabilities of the decisions are significantly different, then the circuit has distinguished F_n from H_n. The general scheme of distinguishing circuits for function generators is shown in Figure 4.2. Goldreich, Goldwasser and Micali were able to construct a

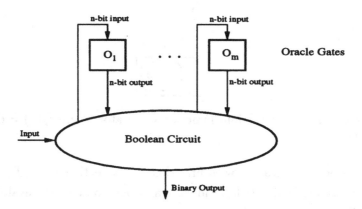

Figure 4.2: The General Scheme of Distinguishers for Function Generators

pseudorandom function generator, given a pseudorandom bit generator which stretched an n-bit seed to a $2n$-bit string [Goldreich *et al.*, 1986]. Their construction is as follows. For a given index x and a given argument y, $f_x(y)$ can be obtained by applying the pseudorandom bit generator n times. The function f_x can be represented as a tree, its lowest layer provides the values of the function and the path specifies the argument of the function. To describe their construction specifically, for $x \in \Sigma^n$ consider

$$f_x(y) = G_y(x)$$

where $G(x) = b_1^x \ldots b_{2n}^x$ is the output of the bit generator for seed x, and $G_i(x)$ is defined recursively as follows:

$$
\begin{aligned}
G_0(x) &= b_1^x \ldots b_n^x \\
G_1(x) &= b_n^x \ldots b_{2n}^x \\
G_y(x) &= G_{y_n}(\ldots(G_{y_2}(G_{y_1}(x)))\ldots)
\end{aligned}
$$

and $y = y_1 y_2 \ldots y_n$. Figure 4.3 shows these operations in a diagram. Goldreich,

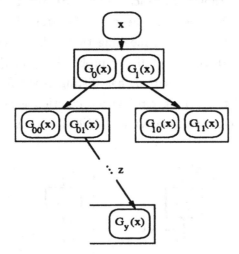

Figure 4.3: A Pseudorandom Function Generator where $f_x(y) = G_y(x)$ with $y = 01z$

Goldwasser and Micali showed first that the collection $F = \{F_n\}$ is a function generator, as it satisfies indexing and polynomial time evaluation, and secondly, that it is pseudorandom.

According to the definition that we gave earlier in Section 4.5 for functions, a function is not necessarily one-to-one. Hence, it is not necessarily invertible. We call those functions that are one-to-one, permutations. Pseudorandom function generators have many applications, but unless they are also invertible, they cannot be used directly in a block cipher cryptosystem. For a time, it was questionble whether it was possible to build pseudorandom invertible permutation generators using pseudorandom function generators.

Luby and Rackoff showed that it is possible to build an invertible pseudorandom permutation generator from three pseudorandom function generators
[Luby and Rackoff, 1988]. In the next section, we explain more about pseudorandom permutation generators and the structure that Luby and Rackoff
put forward.

4.6 Pseudorandom Permutation Generators

4.6.1 Construction

Consider the well known DES cryptographic algorithm. It consists of 16
rounds, where each round is called a Feistel type permutation or a DES-like
permutation. The following gives the precise definition of such a permutation
(an illustration is given in Figure 4.4).

Definition 4.11 *For a function $f \in H_n$, the DES-like permutation associated with f is $D_{2n,f} \in P_{2n}$, defined as*

$$D_{2n,f}(L \parallel R) = (R \oplus f(L) \parallel L)$$

where R and L are n-bit strings, that is, R and L are contained in Σ^n.

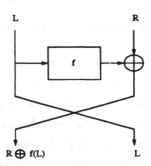

Figure 4.4: A Feistel-type or a DES-like Permutation

Note that, no matter whether f is one-to-one or not, the transformation D is
a permutation. If the above structure incorporates collections of functions at
f, then a collection of invertible permutations would result. If the collection
of functions is a function generator, then the collection of permutations is a

permutation generator. However, the resulting permutation generator is not pseudorandom. This collection can always be distinguished from a collection of random permutations, since the right half of the output is always equal to the left half of the input. A natural question is whether the composition of such permutations would yield a stronger structure.

Definition 4.12 *Having a sequence of functions* $f_1, f_2, \ldots, f_i \in H_n$, *we define the composition of their DES-like permutations as* $\psi \in P_{2n}$, *where*

$$\psi(f_i, \ldots, f_2, f_1) = D_{2n,f_i} \circ D_{2n,f_{i-1}} \circ \ldots \circ D_{2n,f_1}$$

Consider a simple composition $\psi(g, f)$, where the input is $(L \parallel R)$ and the output is $(S \parallel T)$ (see Figure 4.5). If the structure incorporated each of

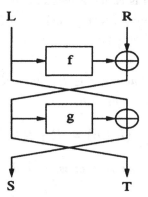

Figure 4.5: Permutation Generator $\psi(g, f)$

a collection of functions at f and g, then a collection of permutations would result. However, this collection is not pseudorandom as there is a circuit given by Luby and Rackoff that is able to distinguish these permutations from a permutation selected randomly from the set of all permutations. The structure of this distinguishing circuit is shown in Figure 4.6, where two oracles are examined with different inputs $L \parallel R_1$ and $L \parallel R_2$. If the oracles evaluate a permutation with a $\psi(g, f)$ structure, then $R_1 \oplus R_2$ is always equal to $T_1 \oplus T_2$. If the permutation is chosen randomly from the set of all permutations, the probability of equality is $\frac{1}{2^n}$. A method for constructing pseudorandom permutation generators from pseudorandom functions, using a DES-like structure, was first presented by Luby and Rackoff. The structure consists of a three-layer composition of DES-like permutations with a

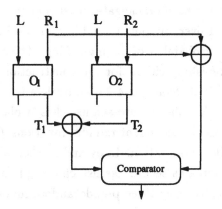

Figure 4.6: A Distinguishing Circuit for $\psi(g, f)$

different pseudorandom function generator at each layer. This structure is shown in Figure 4.7. The following lemma describes the proposed structure

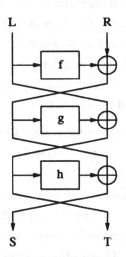

Figure 4.7: Luby and Rackoff's Proposed Structure

and is due to Luby and Rackoff [Luby and Rackoff, 1988].

Lemma 4.1 *Let $f_1, f_2, f_3 \in_r H_n$ be independent random functions and C_{2n} be an oracle circuit with $m < 2^n$ oracle gates; then*

$$| \operatorname{Prob}\{C_{2n}[P_{2n}] = 1\} - \operatorname{Prob}\{C_{2n}[\psi(f_3, f_2, f_1)] = 1\} | \leq \frac{m^2}{2^n}$$

As in practice, the number m of oracle gates used is at most a polynomial in n, then $\frac{m^2}{2^n}$ is less than 1 over any polynomial in n. Note that $\frac{m^2}{2^n}$ is actually an upper bound on the probability of distinguishing. The above lemma says that there is no distinguishing circuit for the construction. It is clear that any distinguishing circuit of Definition 4.9 is equivalent to a chosen Plaintext attack. Therefore a block cipher secure against chosen plaintext attack can be constructed using three independent random functions f_1, f_2 and f_3, and three rounds of DES-like permutation. Luby and Rackoff also demonstrated that the construction remains secure against chosen plaintext attack even when functions are selected from three pseudorandom function generators.

4.6.2   Improvements and Implications

This result was considered a breakthrough in the theory of pseudorandomness, with many cryptographic implications. The proof of the above lemma is based on the assumption that the function used in each layer is a randomly chosen function. Luby and Rackoff considered their result as a justification for the application of DES-like permutations in the design of DES, in the sense that the structure used in DES is sound, although the S-boxes and the functions applied at each round of DES are by no means random.

The result achieved by Luby and Rackoff has attracted much attention to their structures, and since then there have been many researchers trying to improve this result or to apply it for the construction of locally random bit generators and function generators. One such example is by Schnorr, where a construction for locally pseudorandom bit generators is suggested [Schnorr, 1988]. Schnorr proposed using a single pseudorandom function generator f (instead of three), i.e. $\psi(f, f, f)$, to obtain a pseudorandom permutation generator so the amount of necessary memory would be minimized. Then the permutation generator would be used to construct a pseudorandom string generator which stretched $n2^n$ bits to $2n2^{2n}$ bits. Although the construction for the pseudorandom string generator is valid, it was shown by Rueppel that the claim of pseudorandomness for $\psi(f, f, f)$ is not true [Rueppel, 1990]. The distinguishing circuit that he suggested for this permutation generator is shown in Figure 4.8 and is described here. The distinguishing circuit has two oracle gates. A $2n$-bit string $(L \parallel R) \in \Sigma^{2n}$ is fed to the first oracle where the output is $(S_1 \parallel T_1)$. Then the second oracle is fed with $(T_1 \parallel S_1)$, where the output is denoted by $(S_2 \parallel T_2)$. If the permutation used in the evaluation of the oracles has a $\psi(f, f, f)$ structure, $(S_2 \parallel T_2)$ is always equal to $(L \parallel R)$. If the permutation is chosen randomly from the set of all permutations, the probability of equality is $\frac{1}{2^n}$. However, the question whether a smaller number

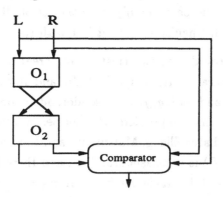

Figure 4.8: A Distinguishing Circuit for $\psi(f, f, f)$

of independent pseudorandom function generators would suffice was still an open problem. Rueppel also showed that $\psi(f, g, f)$ is not pseudorandom. It can be distinguished with the same distinguishing circuit as for $\psi(f, f, f)$. This result[1] was independently obtained by Ohnishi in [Ohnishi, 1988]. He also generalized this result to show that both

$$\psi(f_s, \ldots, f_2, f_1, f_2, \ldots, f_s)$$

and

$$\psi(f_s, \ldots, f_2, f_1, f_1, f_2, \ldots, f_s)$$

are not pseudorandom, where $f_i \in H_n$ for $i = 1 \ldots s$. The description of the distinguishing circuit is as follows:

1. Choose $(L \parallel R) \in \Sigma^{2n}$.

2. Input $(L \parallel R)$ to the first oracle gate O_1. Denote the output of O_1 by $(S_1 \parallel T_1)$.

3. Input $(T_1 \parallel S_1)$ to the second oracle gate O_2. Denote the output of O_2 by $(S_2 \parallel T_2)$.

---

[1] The result was reported in [Zheng *et al.*, 1990c].

4. The distinguisher decides that the permutation generator examined is not pseudorandom if $(L \parallel R) = (S_2 \parallel T_2)$, and outputs a bit **1**.

If a permutation with the structure $\psi(f_s, \ldots, f_2, f_1, f_2, \ldots, f_s)$ is used for the evaluation of the oracles, the output of the above circuit is always **1**. If the permutation is chosen randomly from the set of all permutations, the output of the above distinguishing circuit is **1** with the probability $\frac{1}{2^n}$.

Ohnishi also proved that two (instead of three) independent pseudo-random function generators is enough in Luby and Rackoff's construction, i.e. both $\psi(g, f, f)$ and $\psi(f, g, g)$ are pseudorandom permutation genera-tors. However, it was an open problem whether permutations like $\psi(f^2, f, f)$ were pseudorandom. Later Zheng, Matsumoto and Imai showed that for any $i, j, k \in N$, $\psi(f^k, f^j, f^i)$ is not pseudorandom, and there is a distinguishing circuit with $(m_1 + m_2 + 1)$ oracle gates, where $m_1 = \frac{(i+j)}{d}$, $m_2 = \frac{(j+k)}{d}$ and $d = \gcd(i+j, j+k)$ [Zheng *et al.*, 1990c]. The description of the distinguishing circuit is outlined here.

1. The input to oracle gate O_0 is $(L_0 \parallel R_0) = (0^n \parallel 0^n)$.

2. The input to oracle gate O_1 is $(L_1 \parallel R_1) = (0^n \parallel T_1)$, and if $m_1 > 1$ then for each $1 < p \leq m_1$, the input to oracle gate O_p is $(L_p \parallel R_p) = (0^n \parallel R_{p-1} \oplus S_{p-1})$.

3. The input to oracle gate O_{m+1} is $(L_{m_1+1} \parallel R_{m_1+1}) = (T_0 \parallel 0^n)$, and if $m_2 > 1$, then for each $m_1 + 1 < t \leq m_1 + m_2$, the input to oracle gate O_t is $(L_t \parallel R_t) = (L_{t-1} \oplus T_{t-1} \parallel 0^n)$

4. The distinguisher decides that the permutation generator examined is not pseudorandom if $T_{m_1} = L_{m_1+m_2} \oplus T_{m_1+m_2}$, and outputs a bit **1**.

If the permutation used in the evaluation of the oracles has a $\psi(f^k, f^j, f^i)$ structure, the output of the above circuit is always **1**. If the per-mutation is chosen randomly from the set of all permutations, the output of the above distinguishing circuit is **1** with probability $\frac{1}{2^n}$. The circuit is shown in Figure 4.9. However, the interesting question raised in [Schnorr, 1988] on designing a pseudorandom permutation generator applying only a sin-gle pseudorandom function generator remained unsolved. Finally, Pieprzyk

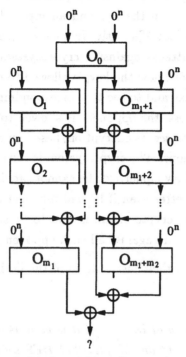

Figure 4.9: A Distinguishing Circuit for $\psi(f^k, f^j, f^i)$

showed that $\psi(f^i, f, f, f)$ for $i \geq 2$ is a pseudorandom permutation genera-
tor [Pieprzyk, 1991]. This result actually solved the open problem raised in
Schnorr's paper and was formally stated by Zheng, Matsumoto and Imai in
[Zheng *et al.*, 1990c].

4.6.3   Security

An encryption algorithm, such as DES, can be considered as a collection or
a family of permutations. For example, in the case of DES, the encryption
algorithm is a collection of 2^{56} permutations, where each permutation is a
member of P_{64} and is indexed by a key k. Similarly, the decryption algorithm
can also be considered as a family of permutations, where the composition of
an encryption algorithm with the corresponding decryption algorithm yields
the identity permutation. In the case of DES, the decryption algorithm is
also a collection of 2^{56} permutations, where each permutation is a member

of P_{64} and is indexed by a key k. Every private-key block cipher should have the property that, given the key and an input, both encryption and decryption can be carried out efficiently. In a chosen plaintext attack, which is one of the strongest attacks against a cryptosystem, it is assumed that an opponent who does not know the key, is allowed to choose a 'reasonable' number of plaintext blocks and to see the corresponding encryption of these blocks. During this process, the opponent is allowed to interactively choose the next plaintext block to see its encryption, based on all previous plaintext blocks and their encryptions. The cryptosystem is considered secure against the opponent if, given a new ciphertext, he cannot predict the corresponding plaintext 'significantly' better than if he had not seen the previous pairs of plaintext-ciphertext. The cryptosystem is said to be secure against chosen plaintext attack if it is secure against all such opponents.

Luby and Rackoff in their justification of the design structure of DES wrote:

The apparent security of DES when it is used as a block private-key cryptosystem rests on the fact that DES seems to pass the black box test, which was informally suggested by Turing. The black box test is the following.

Say that we have two black boxes, one of which computes a fixed randomly chosen function from F_{64} and the other computes DES_k for a fixed randomly chosen k. Then no algorithm which examines the boxes by feeding inputs to them and looking at the outputs can obtain, in a reasonable time, any significant idea about which box is which.

If DES passes the black box test, then it is secure against a chosen plaintext attack when used as a block private key cryptosystem.

Then they added, that it is sufficient that a permutation generator be pseudorandom, to be secure against chosen plaintext attack.

We observed that a permutation generator is pseudorandom if there is no distinguishing circuit, where a distinguishing circuit represents a probabilistic algorithm which has access to a polynomial (in the length of input) number of *input-output* samples via oracle gates. As the oracle gates use permutations only in their normal direction, the distinguishing circuit can choose only inputs and ask for the corresponding outputs. The other notion which was introduced in [Luby and Rackoff, 1988] was that of super-pseudorandomness, which is a stronger property than pseudorandomness. A permutation generator is super-pseudorandom if there is no distinguishing circuit for it, where distinguishing circuits are equipped with both normal oracle gates normal oracle gates and inverse oracle gates. In this case, the distinguishing circuit may choose not only inputs and ask for the corresponding outputs, but may also choose outputs and ask for the corresponding inputs. Such circuits are called super-distinguishing circuits. We give the formal definition for such circuits in the next chapter.

In a chosen plaintext/ciphertext attack, which is an even stronger attack than chosen plaintext attack, an opponent can interactively choose plaintext blocks and see their encryptions and choose ciphertext and see their corresponding plaintext blocks. Thus the opponent is allowed to attack the cryptosystem from 'both ends'. The cryptosystem is considered secure against chosen ciphertext/plaintext attack if, when the opponent is given a new ciphertext, he cannot predict the corresponding plaintext any better than if he had not seen the previous pairs of plaintext-ciphertext. Similarly, if a cryptosystem is secure against chosen ciphertext/plaintext attack, when the opponent is given a new plaintext he cannot predict the corresponding ciphertext any better than if he had not seen the previous pairs of plaintext-ciphertext.

As we mentioned earlier, for a permutation generator to be super-pseudorandom, it should be evaluated with distinguishing circuits equipped with both normal and inverse oracle gates. We should add here that, for a permutation generator, being super-pseudorandom is equivalent to being secure against a chosen plaintext/ciphertext attack. As the meet-in-the-middle attack against a block-cipher-based hash scheme is an oracle circuit containing Boolean gates and has two types of oracle gates that are encryption and decryption, it is a super-distinguishing circuit for the underlying block cipher such that it outputs a bit **1** if two colliding messages are found. Thus meet-in-the-middle attack can be considered a version of chosen plaintext/ciphertext attack against the underlying block cipher. If a block cipher is secure against chosen plaintext/ciphertext attack, the meet-in-the-middle attack cannot successfully be applied against a corresponding block-cipher-based hash scheme. Hence we are interested in developing a structure which can be used for the construction of cipher systems secure against chosen plaintext/ciphertext attacks, such that it can be used for the construction of block-cipher-based hashing algorithms.

Note that every super-pseudorandom permutation generator is also a pseudorandom permutation generator, but it can be shown that the converse is not necessarily true. As an example, it was shown in [Luby and Rackoff, 1988] that although $\psi(h, g, f)$ is a pseudorandom permutation generator, it is not a super-pseudorandom permutation generator. A distinguishing circuit with normal and inverse oracle gates for $\psi(h, g, f)$ can be described as follows. The distinguisher has two normal oracle gates. The first normal oracle gate is fed with $(L \parallel R_1)$ and the second normal oracle gate with $(L \parallel R_2)$, where $R_1 \neq R_2$. Let $(S_1 \parallel T_1)$ and $(S_2 \parallel T_2)$ be the outputs of these two normal oracle gates, respectively. The distinguisher has also an inverse oracle gate with input $(S_2 \oplus R_1 \oplus R_2 \parallel T_2)$. If the last n bits of this inverse oracle gate are equal to $(L \oplus T_1 \oplus T_2)$, the distinguisher decides that the permutation generator examined is not super-pseudorandom, and outputs a bit **1**. If the permutation used in the evaluation of the oracles has a $\psi(h, g, f)$ structure, the output of the above circuit is always **1**. If the permutation is chosen randomly from the set of all permutations, the output of the above super-distinguishing circuit is **1** with probability $\frac{1}{2^n}$. The circuit is shown in Figure 4.10.

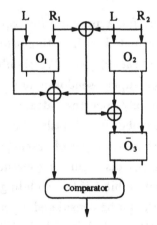

Figure 4.10: A Super-distinguishing Circuit for $\psi(h, g, f)$

4.7   Conclusions

In this chapter, the concepts of indistinguishability and pseudorandomness were presented. It was explained that pseudorandomness of a permutation generator, such as a block cipher, implies its security against chosen plaintext attack. We also explained that, in block-cipher-based hash schemes, we should apply a block cipher which is secure against chosen plaintext/ciphertext attack in order to obtain security against the meet-in-the-middle attack, as such an attack could be transformed into a version of chosen plaintext/ciphertext attack against the underlying block cipher.

It is worthy of note that if a block cipher which is secure against chosen plaintext/ciphertext attack is used in the construction of a hash scheme, the hash scheme need not be collision free. Lai and Massey showed that there may be attacks on the block-cipher-based hash scheme that are easier than attacks on the underlying block cipher alone [Lai and Massey, 1992].

Anyway, Luby and Rackoff's construction of a pseudorandom permutation generator with three rounds of DES-like permutations and three independent pseudorandom function generators and their justification of DES structure based on this result raise the question of how to construct super-pseudorandom permutation generators for use in the construction of stronger block ciphers. Luby and Rackoff proved that $\psi(k, h, g, f)$, a construction with four rounds of DES-like permutations with four independent pseudorandom function generators, yields a super-pseudorandom permutation generator. This result suggests that more rounds should be added to a block cipher secure against chosen plaintext attack to make it resistant to stronger attacks such as chosen plaintext/ciphertext attack. But it does not offer more insight into the construction of a block cipher with a stronger structure. In the next chapter we study super-pseudorandom permutation generators, and we investigate necessary and sufficient conditions for the construction of such generators.

Chapter 5

Construction of Super-Pseudorandom Permutations

5.1 Introduction

In the previous chapter we showed how Luby and Rackoff constructed a pseudorandom invertible permutation generator using three pseudorandom function generators and three rounds of DES-like permutations. This structure is denoted by $\psi(h, g, f)$. Later Pieprzyk showed that four rounds of DES-like permutations with a single pseudorandom function generator $\psi(f^2, f, f, f)$ is also pseudorandom and is secure against a chosen plaintext attack [Pieprzyk, 1991]. Luby and Rackoff also introduced the notion of super-pseudorandomness, where the block cryptosystem is secure against a chosen plaintext/ciphertext attack. They proved that $\psi(h, g, f, e)$ is super-pseudorandom. It remained to be shown how to construct super-pseudorandom permutations and ascertain whether $\psi(f^2, f, f, f)$ was super-pseudorandom.

In this chapter, we present necessary and sufficient conditions for the super-pseudorandomness of DES-like permutations. We further show that four rounds of such permutations with a single random function is not super-pseudorandom. We present a super distinguishing circuit for $\psi(f^2, f, f, f)$ and another one for some cases of $\psi(f^l, f^k, f^j, f^i)$.

Feistel-type permutations were presented in the previous chapter. In this chapter, three generalizations of this class of permutations are presented, where they are called type-1, type-2 and type-3 Feistel transformations. At the end of this chapter, we also investigate the necessary and sufficient conditions for super-pseudorandomness of type-1 Feistel transformations, and we show that using k^2 rounds of this transformation yields a super-pseudorandom permutation generator, where k is the number of branches of the structure. We also show that using $k^2 - k + 1$ rounds of the inverse of this type of transformation is a pseudorandom permutation generator.

The results of this chapter have also appeared in [Sadeghiyan and Pieprzyk, 1991b].

5.2   Super-Pseudorandom Permutations

This section provides some preliminary definitions and notions which are used in this chapter and Chapter 6.

As we mentioned in Chapter 4, the first construction of pseudorandom permutations from pseudorandom functions was presented by Luby and Rackoff. They showed that a block cryptosystem can be constructed which is secure against a chosen plaintext attack when a cryptanalyst can ask for only a polynomial number of plaintext. However, for some cryptographic applications, such as block cipher based hash functions, we require stronger properties for security against a chosen plaintext/ciphertext attack. When the block cryptosystem is secure against the chosen plaintext/ciphertext attack, it is called super-pseudorandom. This notion only applies to invertible permutation generators and is stated formally in the following three definitions.

Definition 5.1 *A permutation generator F is a function generator such that each function $f_{n,k}$ is one-to-one and onto. Let $\overline{F} = \{\overline{F_n} : n \in N\}$, where $\overline{F_n} = \{\overline{f}_{n,k} : k \in \Sigma^{l(n)}\}$, where $\overline{f}_{n,k}$ is the inverse of $f_{n,k}$. F is called invertible if $\overline{F}$ is also a permutation generator.*

Definition 5.2 *A super-distinguishing family of circuits for an invertible permutation generator F is an infinite family of circuits $\{SC_{n_1}, SC_{n_2}, \ldots\}$,* *where $n_1 < n_2 < \ldots$, where each circuit is an oracle circuit containing two types of oracle gates, normal and inverse, such that for some pair of constants c_1 and c_2 and for each $n \in \{n_1, n_2, \ldots\}$ there exist a circuit SC_n with the following properties.*

- *The size of SC_n is less than or equal to n^{c_1}.*

- *If $\mathrm{Prob}\{SC_n[P_n] = 1\}$ is the probability that the output bit of SC_n is one when a permutation p is randomly selected from P_n and p and $\overline{p}$ are used to evaluate normal and inverse oracle gates, and if $\mathrm{Prob}\{SC_n[F_n] = 1\}$ is the probability that the output bit of SC_n is one when a key k of length $l(n)$ is randomly chosen and $f_{n,k}$ and $\overline{f}_{n,k}$ is used to evaluate the normal and inverse oracle gates, respectively, then the distinguishing probability for SC_n is greater than or equal to $\frac{1}{n^{c_2}}$, that is,*

$$| \mathrm{Prob}\{SC_n[P_n] = 1\} - \mathrm{Prob}\{SC_n[F_n] = 1\} | \geq \frac{1}{n^{c_2}}$$

Definition 5.3 *A permutation generator F is super-pseudorandom if there is no super-distinguishing circuit family for F.*

If F is a super-pseudorandom permutation generator, it is secure against the chosen plaintext/ciphertext attack where a cryptanalyst can interactively choose plaintext blocks and view their corresponding cryptograms and also select cryptograms and see their corresponding plaintext blocks.

5.3   Necessary and Sufficient Conditions

It is possible to make a super-pseudorandom permutation generator with four independent random functions, if $f_1, f_2, f_3, f_4 \in H_n$ are independent random functions, then $\psi(f_4, f_3, f_2, f_1)$ is a super-pseudorandom permutation[1]. This was shown by Luby and Rackoff in [Luby and Rackoff, 1988] (see Figure 5.1). This proposal implies that, by increasing the number of rounds and the

---

[1]In Chapters 5 and 6, we say a permutation for a permutation generator, a pseudorandom permutation for a pseudorandom permutation generator and a super-pseudorandom permutation for a super-pseudorandom permutation generator for the sake of brevity.

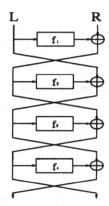

Figure 5.1: A Super-Pseudorandom Permutation Generator

number of pseudorandom functions in a DES-like cryptosystem, better security is attainable. It is question whether a super-pseudorandom permutation
generator can be built with a smaller number of random functions. If it can,
it would then be possible to adopt a sounder structure for the construction
of block ciphers.

In this chapter, we present the necessary and sufficient conditions for
the construction of super-pseudorandom permutations in Theorem 5.1. Next,
these conditions are applied to construct a super-pseudorandom permutation
with fewer random functions. Next, it is shown that a super-pseudorandom
permutation cannot be constructed with a single random function and four
rounds of DES-like permutations.

First a definition for independent permutations is given; this is used
later in this chapter for the proof of Theorem 5.1.

Definition 5.4 *A D-distinguishing family of circuits for two invertible pseudorandom permutation generators* (Π_1, Π_2) *is an infinite family of circuits*
$\{DC_{n_1}, DC_{n_2}, \ldots\}$, *where* $n_1 < n_2 < \ldots$, *where each circuit is an oracle circuit containing two types of oracle gates, such that for some pair of constants*
c_1 *and* c_2 *and for each* $n \in \{n_1, n_2, \ldots\}$ *there exists a circuit* DC_n *with the
following properties.*

- *The size of* DC_n *is less than or equal to* n^{c_1}.

- *If* $\mathrm{Prob}\{DC_n[P_n, P_n] = 1\}$ *is the probability that the output bit of* DC_n
  is 1, when two permutations p_1 *and* p_2 *are chosen independently and*

randomly from P_n and are used to evaluate the two types of oracle gates of DC_n, respectively, and if $\mathrm{Prob}\{DC_n[\Pi_1, \Pi_2] = 1\}$ is the probability that the output bit of DC_n is one when a key k of length $l(n)$ is randomly chosen and $p_{1,k} \in \Pi_1$ and $p_{2,k} \in \Pi_2$ are used to evaluate the two types of oracle gates, respectively, then the distinguishing probability for DC_n is greater than or equal to $\frac{1}{n^{c_2}}$, that is,

$$| \mathrm{Prob}\{DC_n[P_n, P_n] = 1\} - \mathrm{Prob}\{DC_n[\Pi_1, \Pi_2] = 1\} | \geq \frac{1}{n^{c_2}}$$

Definition 5.5 *Two pseudorandom permutation generators, Π_1 and Π_2, are said to be independent if there is no D-distinguishing oracle circuit family for (Π_1, Π_2).*

Note that the D-distinguishing oracle circuits are generalizations of the distinguishing circuits and the super-distinguishing circuits if two simple tests are excluded. If there is no distinguishing circuit family for Π_1, then there is no D-distinguishing circuit for the permutation generators Π_1 and Π_1 itself, provided that the D-distinguishing circuit is not testing the identity of the two permutation generators (for example, giving an input to the two types of oracles and comparing the outputs). Moreover, if there is no super-distinguishing circuit family for Π_1, then there is no D-distinguishing circuit for a permutation generator Π_1 and its inverse $\overline{\Pi}_1$, and provided that the D-distinguishing circuit is not testing to see whether the two permutation generators are inverse to each other (for example, giving an input to one type of oracle and feeding the other type of oracle with this result and comparing the output with the original input). Furthermore, the converse of these statements is also true. For example, if there is no D-distinguishing circuit for Π_1 and Π_1 itself, except if the circuit is testing the identity of the two permutation generators, then the permutation generator Π_1 is pseudorandom. We apply D-distinguishing circuits as a tool for the evaluation of both pseudorandomness and super-pseudorandomness.

The following lemma shows how to construct two independent permutations, applying DES-like structures.

Lemma 5.1 *Let $f_1, f_2, \ldots, f_i \in_r F_n$, where $i \in N$. Then $G_2 = \psi(f_i, \ldots, f_2)$ and $G_3 = \psi(f_1, \ldots, f_{i-1})$ are two independent permutations if and only if they are pseudorandom.*

Proof : First, we show that if $G_2 = \psi(f_i, \ldots, f_2)$ and $G_3 = \psi(f_1, \ldots, f_{i-1})$, are pseudorandom, then they are independent permutations, where $f_1, \ldots, f_i \in_r F_n$. For simplicity let $f_1, \ldots, f_i \in_r H_n$, and suppose that one is allowed to examine only a polynomial (in n) number of oracle gates. When G_2 and G_3 are pseudorandom, it is assumed that the probability of distinguishing G_2 or G_3 from a random permutation is less than $\frac{1}{n^{c_2}}$, for any constant c_2 and a sufficiently large n. Since both G_2 and G_3 are indistinguishable from random permutations, any distinguishing circuit for the dependency of G_2 and G_3 would be a circuit which estimates the output value at least from either branch of G_2 or G_3 for input, when a polynomial number of G_2 and G_3 oracles are examined. As, both G_2 and G_3 have two branches, two situations may arise.

- The number of rounds i is even.

    When i is even, each branch of G_2 and G_3 is fed with a different set of random functions. Figure 5.2 illustrates these structures. Denote the

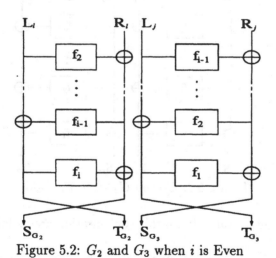

Figure 5.2: G_2 and G_3 when i is Even

output of G_2 on an input of $(L_l \parallel R_l)$ by $(S_{G_2} \parallel T_{G_2})$ and the output of G_3 on an input of $(L_j \parallel R_j)$ by $(S_{G_3} \parallel T_{G_3})$.

$$
\begin{aligned}
S_{G_2} &= R_l \oplus f_2(L_l) \oplus \ldots \oplus f_{i-2}(.) \oplus f_i(T_{G_2}) \\
T_{G_2} &= L_l \oplus f_3(.) \oplus \ldots \oplus f_{i-1}(.) \\
S_{G_3} &= R_j \oplus f_{i-1}(L_j) \oplus \ldots \oplus f_3(.) \oplus f_1(T_{G_3}) \\
T_{G_3} &= L_j \oplus f_{i-2}(.) \oplus \ldots \oplus f_2(.)
\end{aligned}
$$

As each function $f_1, \ldots, f_i$ is chosen independently and randomly from the set of all functions, the probability that these two random variables are the same in m oracle gates is equal to $\frac{m(m-1)}{2^n}$. As there are four random variables, the probability that two of them are the same is $\frac{6m(m-1)}{2^n}$. When m is polynomial in n, this probability is less than $\frac{1}{n^{c_2}}$, for any constant c_2 and a sufficiently large n. If $f_1, \ldots, f_i$ were chosen from F_n rather from H_n, the probability that two of the above four random variables are the same would remain less than 1 over any polynomial in n. Hence, the above four variables are independent of one another, and there is no D-distinguishing circuit for G_2 and G_3.

- The number of rounds i is odd.

In this structure, one branch of G_2 is fed with the same set of random functions which feeds a branch of G_3, but the other branches are fed with a different set of random functions (See Figure 5.3). The four

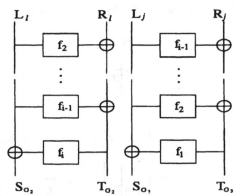

Figure 5.3: G_2 and G_3 when i is Odd

output random variables in this configuration are as follows.

$$
\begin{aligned}
S_{G_2} &= L_l \oplus f_3(.) \oplus \ldots \oplus f_{i-2} \oplus f_i(T_{G_2}) \\
T_{G_2} &= R_l \oplus f_2(L_l) \oplus \ldots \oplus f_{i-1}(.) \\
S_{G_3} &= L_j \oplus f_{i-2}(.) \oplus \ldots \oplus f_1(T_{G_3}) \\
T_{G_3} &= R_j \oplus f_{i-1}(L_l) \oplus f_{i-3} \oplus \ldots \oplus f_2(.)
\end{aligned}
$$

T_{G_2} and T_{G_3} are the sum of outputs of the same set of random functions, but with a reverse ordering. In general, when two functions f and g are chosen independently and randomly from H_n, $(f \circ g)(x)$ is independent of $(g \circ f)(x)$ for any x. As the functions $f_1, \ldots, f_i$ are chosen independently and randomly from H_n, then the probability that two random variables T_{G_2} and T_{G_3} are the same in m oracle gates is $\frac{m(m-1)}{2^n}$. Hence, the probability that two of the above four random variables are the same is $\frac{6m(m-1)}{2^n}$. Again, when m is polynomial in n, this probability is less than $\frac{1}{n^{c_2}}$, for any constant c_2 and a sufficiently large n. When random functions are replaced by pseudorandom ones, the probability of dependency between G_2 and G_3 remains less than 1 over any polynomial in n (see [Luby and Rackoff, 1988].

Although, in the above proof, we assumed that the pseudorandom functions $f_1, \ldots, f_i$ are chosen independently from F_n, this is not a necessary condition. Generally, it is sufficient that only f_{i-1} and f_2 be chosen independently to make the output random variables of the two pseudorandom permutation generators G_2 and G_3 independent, and it does not matter whether the other pseudorandom functions are chosen independently.

We next show that, if G_2 and G_3 are independent, then they are pseudorandom. If G_2 and G_3 are independent, there is no oracle circuit equipped with two types of oracle gates to distinguish them from two independently chosen random functions. In other words, the probability that such a family of oracle circuits distinguishes G_2 and G_3 from two independently chosen random functions is not greater than (or equal to) $\frac{1}{n^{c_2}}$ for some constant c_2 and for each n. As an oracle circuit with two types of oracle gates is a much stronger distinguisher than an oracle circuit with only one type of oracle gate, using only one type of oracle reduces the possibility of distinguishing. Hence, the following relations hold.

$$| \operatorname{Prob}\{C_{2n}[G_2] = 1\} - \operatorname{Prob}\{C_{2n}[P_{2n}] = 1\} | \leq$$
$$| \operatorname{Prob}\{DC_{2n}[G_2, G_3] = 1\} - \operatorname{Prob}\{DC_{2n}[P_{2n}, P_{2n}] = 1\} | < \frac{1}{n^{c_2}}$$

and also

$$| \operatorname{Prob}\{C_{2n}[G_3] = 1\} - \operatorname{Prob}\{C_{2n}[P_{2n}] = 1\} | \leq$$
$$| \operatorname{Prob}\{DC_{2n}[G_2, G_3] = 1\} - \operatorname{Prob}\{DC_{2n}[P_{2n}, P_{2n}] = 1\} | < \frac{1}{n^{c_2}}$$

The above inequalities show that both G_2 and G_3 are pseudorandom when they are independent. This completes the proof of Lemma 5.1.   □

Theorem 5.1 *Let $f_1, f_2, \ldots, f_i \in F_n$ such that $G_1 = \psi(f_i, \ldots, f_1)$ is pseudorandom. Then G_1 is super-pseudorandom if and only if $G_2 = \psi(f_i, \ldots, f_2)$ and $G_3 = \psi(f_1, \ldots, f_{i-1})$ are independent permutations.*

Proof :   To prove Theorem 5.1, we prove two lemmas A and B.

- **Lemma A** *If G_2 and G_3 are independent, then G_1 is super-pseudorandom.*

To prove this, it is necessary to show that $G_3 = \psi(f_1, \ldots, f_{i-1})$ and $\overline{G}_1 = \psi(f_1, \ldots, f_{i-1}, f_i)$ are independent of each other. Figure 5.4 shows these structures with respect to each other. The validity of this claim can be

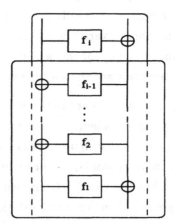

Figure 5.4: $\overline{G}_1$ and G_3 with Respect to Each Other

shown by contradiction. Assume that $\overline{G}_1$ and G_3 are not independent; there exists a D-distinguishing circuit family for which

$$| \text{Prob}\{DC_{2n}[G_3, \overline{G}_1] = 1\} - \text{Prob}\{DC_{2n}[P_{2n}, P_{2n}] = 1\} | \geq \frac{1}{n^{c_2}}$$

for some constant c_2. Without changing the inequality relation, we have

$$| \text{Prob}\{DC_{2n}[G_3, \overline{G}_1] = 1\} - \text{Prob}\{DC_{2n}[G_3, G_3] = 1\} +$$
$$\text{Prob}\{DC_{2n}[G_3, G_3] = 1\} - \text{Prob}\{DC_{2n}[P_{2n}, P_{2n}] = 1\} | \geq \frac{1}{n^{c_2}}$$

Then

$$| \text{Prob}\{DC_{2n}[G_3, \overline{G}_1] = 1\} - \text{Prob}\{DC_{2n}[G_3, G_3] = 1\} | +$$
$$| \text{Prob}\{DC_{2n}[G_3, G_3] = 1\} - \text{Prob}\{DC_{2n}[P_{2n}, P_{2n}] = 1\} | \geq \frac{1}{n^{c_2}}$$

If $\mid \text{Prob}\{DC_{2n}[G_3, G_3] = 1\} - \text{Prob}\{DC_{2n}[P_{2n}, P_{2n}] = 1\} \mid \geq \frac{1}{n^c}$ for some constant c, then G_3 is not pseudorandom; this contradicts our assumption. If $\mid \text{Prob}\{DC_{2n}[G_3, \overline{G}_1] = 1\} - \text{Prob}\{DC_{2n}[G_3, G_3] = 1\} \mid \geq \frac{1}{n^c}$ for some constant c, then the oracle circuit distinguishes f_i from a randomly chosen function. This also contradicts our assumption that f_i is chosen from a pseudorandom function generator. Since both cases lead to contradictions, we conclude that G_3 and $\overline{G}_1$ are independent of each other. Note that, in order for $\overline{G}_1$ and G_3 to be two independent permutations, there is no need that the pseudorandom function f_i be chosen independently of $f_1, \ldots, f_{i-1}$.

By Lemma A, G_2 and G_3 are independent. Hence

$$\mid \text{Prob}\{DC_{2n}[G_2, G_3] = 1\} - \text{Prob}\{DC_{2n}[P_{2n}, P_{2n}] = 1\} \mid < \frac{1}{n^{c_2}}$$

for any constant c_2. Without changing the sign of inequality, we may rewrite the above relation as,

$$\mid \text{Prob}\{DC_{2n}[G_1, \overline{G}_1] = 1\} - \text{Prob}\{DC_{2n}[G_1, \overline{G}_1] = 1\} \; +$$
$$\text{Prob}\{DC_{2n}[G_3, \overline{G}_1] = 1\} - \text{Prob}\{DC_{2n}[G_3, \overline{G}_1] = 1\} \; +$$
$$\text{Prob}\{DC_{2n}[G_2, G_3] = 1\} - \text{Prob}\{DC_{2n}[P_{2n}, P_{2n}] = 1\} \mid \; < \; \frac{1}{n^{c_2}}$$

Then

$$\mid\mid \text{Prob}\{DC_{2n}[G_2, G_3] = 1\} - \text{Prob}\{DC_{2n}[G_3, \overline{G}_1] = 1\} \mid \; -$$
$$\mid \text{Prob}\{DC_{2n}[G_3, \overline{G}_1] = 1\} - \text{Prob}\{DC_{2n}[G_1, \overline{G}_1] = 1\} \mid \; -$$
$$\mid \text{Prob}\{DC_{2n}[G_1, \overline{G}_1] = 1\} - \text{Prob}\{DC_{2n}[P_{2n}, P_{2n}] = 1\} \mid\mid \; < \; \frac{1}{n^{c_2}}$$

Since it was assumed that G_2 and G_3 are two independent permutations, and so are G_3 and $\overline{G}_1$, then $\mid \text{Prob}\{DC_{2n}[G_2, G_3] = 1\} - \text{Prob}\{DC_{2n}[G_3, \overline{G}_1] = 1\} \mid$ is less than $\frac{1}{n^c}$ for any constant c, since

$$\mid \text{Prob}\{DC_{2n}[G_2, G_3] = 1\} - \text{Prob}\{DC_{2n}[G_3, \overline{G}_1] = 1\} \mid \; <$$
$$\mid \text{Prob}\{DC_{2n}[G_2, G_3] = 1\} - \text{Prob}\{DC_{2n}[P_{2n}, P_{2n}] = 1\} \mid \; +$$
$$\mid \text{Prob}\{DC_{2n}[G_3, \overline{G}_1] = 1\} - \text{Prob}\{DC_{2n}[P_{2n}, P_{2n}] = 1\} \mid \; < \; \frac{1}{n^c}$$

Thus

$$\mid \text{Prob}\{DC_{2n}[G_1, \overline{G}_1] = 1\} - \text{Prob}\{DC_{2n}[P_{2n}, P_{2n}] = 1\} \mid \; +$$
$$\mid \text{Prob}\{DC_{2n}[G_3, \overline{G}_1] = 1\} - \text{Prob}\{DC_{2n}[G_1, \overline{G}_1] = 1\} \mid \; < \; \frac{1}{n^{c_2}}$$

Hence, each of the above absolute values is less than $\frac{1}{n^{c_2}}$. In other words

$$| \operatorname{Prob}\{DC_{2n}[G_1, \overline{G}_1] = 1\} - \operatorname{Prob}\{DC_{2n}[P_{2n}, P_{2n}] = 1\} | < \frac{1}{n^{c_2}}$$

Hence G_1 and $\overline{G}_1$ are independent of each other, and G_1 is a super-pseudo-random permutation, as it is pseudorandom.

To conclude Theorem 5.1, we also need Lemma B.

- **Lemma B** *If G_1 is a super-pseudorandom permutation, then G_2 and G_3 are two independent permutations.*

According to the assumption of this lemma, we have that

$$| \operatorname{Prob}\{DC_{2n}[G_1, \overline{G}_1] = 1\} - \operatorname{Prob}\{DC_{2n}[P_{2n}, P_{2n}] = 1\} | < \frac{1}{n^{c_2}}$$

for any constant c_2. Without changing the sign of inequality,

$$\begin{aligned}
| \operatorname{Prob}\{DC_{2n}[G_1, \overline{G}_1] = 1\} - \operatorname{Prob}\{DC_{2n}[G_3, \overline{G}_1] = 1\} \ + \\
\operatorname{Prob}\{DC_{2n}[G_3, \overline{G}_1] = 1\} - \operatorname{Prob}\{DC_{2n}[G_3, G_2] = 1\} \ + \\
\operatorname{Prob}\{DC_{2n}[G_3, G_2] = 1\} - \operatorname{Prob}\{DC_{2n}[P_{2n}, P_{2n}] = 1\} | \ < \ \frac{1}{n^{c_2}}
\end{aligned}$$

Then

$$\begin{aligned}
|| \operatorname{Prob}\{DC_{2n}[G_1, \overline{G}_1] = 1\} - \operatorname{Prob}\{DC_{2n}[G_3, \overline{G}_1] = 1\} | \ - \\
| \operatorname{Prob}\{DC_{2n}[G_3, \overline{G}_1] = 1\} - \operatorname{Prob}\{DC_{2n}[G_3, G_2] = 1\} | \ - \\
| \operatorname{Prob}\{DC_{2n}[G_3, G_2] = 1\} - \operatorname{Prob}\{DC_{2n}[P_{2n}, P_{2n}] = 1\} || \ < \ \frac{1}{n^{c_2}}
\end{aligned}$$

If either of G_2 or G_3 are not pseudorandom, it can be shown that G_1 is not super-pseudorandom. To justify this claim, suppose there is a probabilistic distinguishing circuit with m oracle gates which distinguishes G_3 from a random permutation with a probability better than $\frac{1}{n^c}$ for some constant c. In other words, when m inputs, i.e. $S_j \parallel T_j$, are fed to its (only normal) oracle gates, the probability of obtaining a desired output is bigger than $\frac{1}{n^c}$. Suppose that a $2n$ bit string $(R_i \parallel L_i)$ is fed to a normal oracle gate of G_1 and the output is $(S_i \parallel T_i)$. If $(R_i \parallel L_i)$ is fed to $\overline{G}_3$, the output is either $(T_i \parallel S_i')$ or $(T_i' \parallel S_i)$, depending on whether $\overline{G}_3$ consists of an odd or an even number of rounds. Without loss of generality, assume that the output

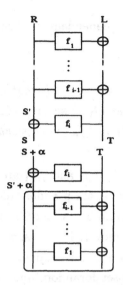

Figure 5.5: When an Inverse and a Normal Gate are Applied with Each Other

is $(T_i \parallel S'_i)$. Now, if $(S_i \oplus \alpha \parallel T_i)$ is fed to an inverse oracle gate of G_1, the output is equivalent to the output of a normal oracle gate, G_3, when it is fed with $(S'_i \oplus \alpha \parallel T_i)$. This procedure is depicted in Figure 5.5. Hence, it is possible to obtain a desired output with a probability better than $\frac{1}{n^c}$, when m inverse oracle gates of G_1 are examined with different values for α. Then a probabilistic super-distinguishing circuit for G_1 with at most m^2 normal oracles and m^2 inverse oracles would be able to yield the same output with the same probability.

Suppose it can be proved that G_3 is pseudorandom; then it can be shown that

$$\mid \mathrm{Prob}\{DC_{2n}[G_1, \overline{G}_1] = 1\} - \mathrm{Prob}\{DC_{2n}[G_3, \overline{G}_1] = 1\} \mid < \frac{1}{n^c}$$

for any constant c, since G_3 and $\overline{G}_1$ are independent of each other (this was proved in Lemma A). As G_1 is assumed to be super-pseudorandom, the following inequality is also valid

$$\mid \mathrm{Prob}\{DC_{2n}[G_3, \overline{G}_1] = 1\} - \mathrm{Prob}\{DC'_{2n}[G_3, G_2] = 1\} \mid +$$
$$\mid \mathrm{Prob}\{DC_{2n}[G_3, G_2] = 1\} - \mathrm{Prob}\{DC_{2n}[P_{2n}, P_{2n}] = 1\} \mid < \frac{1}{n^{c_2}}$$

Hence

$$\mid \mathrm{Prob}\{DC_{2n}[G_3, G_2] = 1\} - \mathrm{Prob}\{DC_{2n}[P_{2n}, P_{2n}] = 1\} \mid < \frac{1}{n^{c_2}}$$

In other words, G_2 and G_3 are independent of each other. This completes the proof of Theorem 5.1.   □

Corollary 5.1 *Let* $f_1, f_2, \ldots, f_i \in_r F_n$ *such that* $G_1 = \psi(f_i, \ldots, f_1)$ *is a pseudorandom permutation. Then* G_1 *is super-pseudorandom if and only if* $G_2 = \psi(f_i, \ldots, f_2)$ *and* $G_3 = \psi(f_1, \ldots, f_{i-1})$ *are pseudorandom permutations.*

Proof : As shown in Lemma 5.1, if f_2 and f_{i-1} are two independent pseudorandom functions, and G_2 and G_3 are pseudorandom, then they are independent. As was shown in Theorem 5.1, when G_2 and G_3 are independent, G_1 is super-pseudorandom. Moreover, it was shown that, if G_1 is super-pseudorandom, then G_2 and G_3 are independent of each other, and when they are independent both are pseudorandom and satisfy the conditions of Theorem 5.1. This finishes the proof.   □

Ohnishi showed that it is possible to use two independent pseudorandom functions, instead of three, in a three round DES-like structure to obtain a pseudorandom permutation. That is, $\psi(f_2, f_2, f_1)$ is a pseudorandom permutation generator [Ohnishi, 1988]. Applying his results and Theorem 5.1, we have the following corollary.

Corollary 5.2 *Let* $f_1, f_2 \in_r F_n$; *then* $G_1 = \psi(f_2, f_2, f_1, f_1)$ *is a super-pseudorandom permutation.*

Later Patarin also proved this corollary, using another method for the evaluation of super-pseudorandomness [Patarin, 1992]. This structure is depicted in Figure 5.6. It turns out that it is possible to construct super-distinguishing circuits with a distinguishing probability near 1 for some structures. We now investigate selected structures and show how to construct super-distinguishing circuits for them.

Lemma 5.2 *Let* $f \in_r H_n$; *then* $\psi(f^2, f, f, f)$ *is not super-pseudorandom and there is a super-distinguishing circuit* SC_{2n} *with 4 normal and inverse oracle gates.*

Proof : By Theorem 5.1, $\psi(f^2, f, f, f)$ is super-pseudorandom if $\psi(f^2, f, f)$ and $\psi(f, f, f)$ are independent. Zheng, Matsumoto and Imai showed that

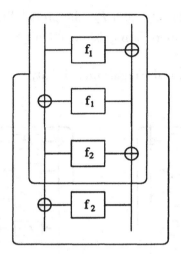

Figure 5.6: A Super-Pseudorandom Permutation Generator with Two Pseudorandom Function Generators

it is impossible to get a pseudorandom permutation with three rounds of DES-like permutations and a single random function [Zheng *et al.*, 1990c]; so neither $\psi(f^2, f, f)$ nor $\psi(f, f, f)$ are pseudorandom, and they are not therefore independent. Therefore $\psi(f^2, f, f, f)$ cannot be a super-pseudorandom permutation by Theorem 5.1. The structure of a super-distinguishing circuit SC_{2n} is as follows.

Let $\hat{O}_0, \hat{O}_1, \hat{O}_3$ be normal oracle gates and let $\check{O}_2$ be an inverse oracle gate. Denote by $(L_u \parallel R_u)$ and $(S_u \parallel T_u)$, respectively, the input to and the output of the u-th oracle gate, and denote by $0^n \in \Sigma^n$ an n-bit string of all **0**.

1. The input to $\hat{O}_0$ is $(L_0 \parallel R_0) = (0^n \parallel 0^n)$

2. The input to $\hat{O}_1$ is $(L_1 \parallel R_1) = (0^n \parallel T_0)$

3. The input to $\check{O}_2$ is $(L_2 \parallel R_2) = (0^n \parallel 0^n)$

4. The input to $\hat{O}_3$ is $(L_3 \parallel R_3) = (S_2 \parallel T_2 \oplus T_0)$

5. SC_{2n} outputs a **1** if and only if $T_3 = T_0 \oplus T_1$

When a function $\psi(f^2, f, f, f)$ is used to evaluate the oracle gates, the probability that SC_{2n} outputs a **1**, is equal to 1, and when a function is drawn randomly and uniformly from P_{2n} the probability that SC_{2n} outputs 1 is $\frac{1}{2^n}$. Thus SC_{2n} is a super-distinguishing circuit for $\psi(f^2, f, f, f)$. $\square$

Figure 5.7 depicts this distinguishing circuit.

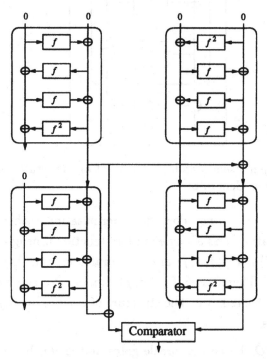

Figure 5.7: A Super-distinguishing Circuit for $\psi(f^2, f, f, f)$

Lemma 5.2 can be generalized to the following theorem.

Theorem 5.2 *Let $f \in_r H_n$; there is a super-distinguishing circuit with $p + q + 4$ normal and inverse oracle gates for $\psi(f^l, f^k, f^j, f^i)$, where p and q satisfy*

$$i + j - l = (q - p)(j + k)$$

Proof : Because of the results of Zheng, Matsumoto and Imai, neither $\psi(f^l, f^k, f^j)$ nor $\psi(f^i, f^j, f^k)$ is pseudorandom [Zheng et al., 1990c]. Thus, by Theorem 5.1, $\psi(f^l, f^k, f^j, f^i)$ is not super-pseudorandom. We have constructed a distinguisher for the following case.

Let $\hat{O}_0, \hat{O}_1, \ldots, \hat{O}_{p+1}$ and $\hat{O}_{p+q+3}$ be normal oracle gates and let $\check{O}_{p+2}$, $\ldots, \check{O}_{p+q+2}$ be inverse oracle gates, where p and q satisfy: $i+j+k+p(j+k) = l + k + q(j+k)$ or,

$$i + j - l = (q - p)(j + k)$$

The structure of SC_{2n} is as follows:

1. The input to $\hat{O}_0$ is $(L_0 \parallel R_0) = (0^n \parallel 0^n)$

2. The input to $\hat{O}_1$ is $(L_1 \parallel R_1) = (0^n \parallel T_0)$ and the input for $\hat{O}_2$ to $\hat{O}_{p+1}$ is $(L_u \parallel R_u) = (0^n \parallel T_{u-1} \oplus R_{u-1})$

3. The input to $\check{O}_{p+2}$ is $(L_{p+2} \parallel R_{p+2}) = (0^n \parallel 0^n)$

4. The input to $\check{O}_{p+3}$ is $(L_{p+3} \parallel R_{p+3}) = (S_{p+3} \parallel 0^n)$ and the input for $\check{O}_{p+4}$ to $\check{O}_{p+q+2}$ is $(L_u \parallel R_u) = (L_{u-1} \oplus T_{u-1} \parallel 0^n)$

5. The input to $\hat{O}_{p+q+3}$ is $(L_{p+q+3} \parallel R_{p+q+3}) = (S_{p+q+2} \parallel T_p \oplus T_{p+q+2})$

6. SC_{2n} outputs a **1** if and only if $T_{p+q+3} = T_p \oplus T_{p+1}$

When a function $\psi(f^l, f^k, f^j, f^i)$ is used to evaluate the oracle gates and $i+j+k+p(j+k) = l+k+q(j+k)$, the probability that C_{2n} outputs a **1**, is equal to 1, and when a function is drawn randomly and uniformly from P_{2n}, the probability that C_{2n} outputs **1**, is equal to $\frac{1}{2^n}$.   □ Figure 5.8

depicts this distinguishing circuit.

5.4 Super-Pseudorandomness in Generalized DES-like Permutations

A DES-like permutation is a permutation in P_{2n} that uses functions in H_n. Zheng, Matsumoto and Imai made three types of permutations in P_{kn} by generalizing the construction of the DES-like permutation and application of functions in H_n, and called them type-1, type-2 and type-3 Feistel transformations [Zheng et al., 1990d]. In this section, we show that k^2 rounds of type-1 transformations are required to get a super-pseudorandom permutation.

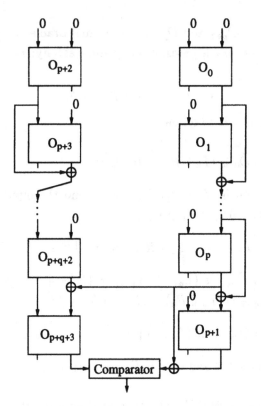

Figure 5.8: A Distinguishing Circuit for $\psi(f^l, f^k, f^j, f^i)$

First, we present the definition of type-1 transformations. Then, the necessary and sufficient conditions for the super-pseudorandomness of this type of transformation will be presented; accordingly, some cases which are pseudorandom but cannot be super pseudorandom, will be given. Finally, we show that k^2 rounds of type-1 transformations produce a super-pseudorandom permutation.

5.4.1 Feistel-Type Transformations

Type-1 Transformations

Let $g_{1,i} \in H_{kn}$ be a function associated with an $f_i \in H_n$ and defined by

$$g_{1,i}(B_1 \parallel B_2 \parallel \cdots \parallel B_k) = (B_2 \oplus f_i(B_1) \parallel B_3 \parallel \cdots \parallel B_k \parallel B_1)$$

where $B_j \in \Sigma^n$ for $1 \leq j \leq k$ and $k \in N$. Functions defined in such a way are called type-1 transformations (See Figure 5.9). $g_{1,i}$ can be decomposed

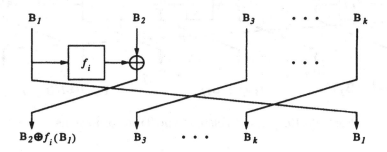

Figure 5.9: Type-1 Feistel Type Transformations

into $g_{1,i} = L_{rot} \circ \pi_{1,i}$, where

$$\pi_{1,i}(B_1 \parallel B_2 \parallel \ldots \parallel B_k) = (B_1 \parallel B_2 \oplus f_i(B_1) \parallel B_3 \parallel \ldots \parallel B_k)$$
$$L_{rot}(B_1 \parallel B_2 \parallel \ldots \parallel B_k) = (B_2 \parallel B_3 \parallel \ldots \parallel B_k \parallel B_1)$$

The function $g_{1,i}$ is an invertible permutation on Σ^{kn}, and its inverse, denoted by $\overline{g}_{1,i}$ is given by $\overline{g}_{1,i} = \pi_{1,i} \circ R_{rot}$, where

$$R_{rot}(B_1 \parallel B_2 \parallel \ldots \parallel B_k) = (B_k \parallel B_1 \parallel B_2 \parallel \ldots \parallel B_{k-1})$$

For $f_1, f_2, \ldots, f_s \in H_n$, define $\psi_1(f_s, \ldots, f_2, f_1) = g_{1,s} \circ \ldots \circ g_{1,2} \circ g_{1,1}$. Note that ψ_1 is an invertible permutation on Σ^{kn}, and its inverse $\overline{\psi}_1$ is defined by

$$\overline{\psi}_1(f_1, f_2, \ldots, f_s) = \overline{g}_{1,1} \circ \overline{g}_{1,2} \circ \ldots \circ \overline{g}_{1,s}$$

Type-2 Transformations

Let $g_{2,i} \in H_{kn}$ be a function associated with a function tuple

$$h_i = (f_{i,1}, f_{i,3}, \ldots, f_{i,2l-1}),$$

where $l = \frac{k}{2}$ and $f_{i,j} \in H_n$, and defined by

$$g_{2,i}(B_1 \parallel B_2 \parallel \ldots \parallel B_k) = (B_2 \oplus f_{i,1}(B_1) \parallel \ldots \parallel B_{k-1} \parallel B_k \oplus f_{i,k-1}(B_{k-1}) \parallel B_1)$$

where $B_j \in \Sigma^n$ for $1 \leq j \leq k$ and $k \in N$ (See Figure 5.10). Functions defined in such a way are called type-2 transformations. $g_{2,i}$ can be decomposed into

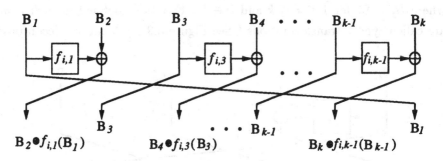

Figure 5.10: Type-2 Feistel-Type Transformations

$g_{2,i} = L_{rot} \circ \pi_{2,i}$, where

$$\pi_{2,i}(B_1 \parallel B_2 \parallel \ldots \parallel B_k) =$$
$$(B_1 \parallel B_2 \oplus f_{i,1}(B_1) \parallel \ldots \parallel B_{k-1} \parallel B_k \oplus f_{i,k-1}(B_{k-1}))$$

The function $g_{2,i}$ is an invertible permutation on Σ^{kn}, and its inverse, denoted by $\overline{g}_{2,i}$ is given by $\overline{g}_{2,i} = \pi_{2,i} \circ R_{rot}$.

For s-tuples of functions $(h_1, h_2, \ldots, h_s)$, define

$$\psi_2(h_s, \ldots, h_2, h_1) = g_{2,s} \circ \ldots \circ g_{2,2} \circ g_{2,1}$$

Note that ψ_2 is an invertible permutation on Σ^{kn}, and its inverse $\overline{\psi}_2$ is defined by

$$\overline{\psi}_2(h_1, h_2, \ldots, h_s) = \overline{g}_{2,1} \circ \overline{g}_{2,2} \circ \ldots \circ \overline{g}_{2,s}$$

Type-3 Transformations

Let $g_{2,i} \in H_{kn}$ be a function associated with a function-tuple

$$h_i = (f_{i,1}, f_{i,3}, \ldots, f_{i,k-1})$$

, where $f_{i,j} \in H_n$. Then $g_{3,i}$ is defined by

$$g_{3,i}(B_1 \parallel B_2 \parallel \ldots \parallel B_k) = (B_2 \oplus f_{i,1}(B_1) \parallel \ldots \parallel B_k \oplus f_{i,k-1}(B_{k-1}) \parallel B_1)$$

where $B_j \in \Sigma^n$ for $1 \leq j \leq k$ and $k \in N$ (See Figure 5.11). Functions defined in such a way are called type-3 transformations. The function $g_{3,i}$ can be decomposed into $g_{3,i} = L_{rot} \circ \pi_{3,i}$, where

$$\pi_{3,i}(B_1 \parallel B_2 \parallel \ldots \parallel B_k) = (B_1 \parallel B_2 \oplus f_{i,1}(B_1) \parallel \ldots \parallel B_k \oplus f_{i,k-1}(B_{k-1}))$$

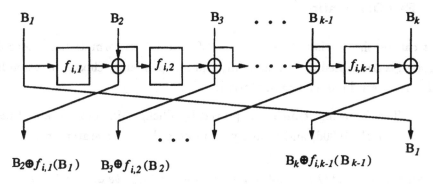

Figure 5.11: Type-3 Feistel-Type Transformations

The function $g_{3,i}$ is an invertible permutation on Σ^{kn}, and its inverse, denoted by $\bar{g}_{3,i}$ is given by $\bar{g}_{3,i} = \bar{\pi}_{3,i} \circ R_{rot}$. Note that $\bar{\pi}_{3,i}(C_1 \parallel \ldots \parallel C_k) = (B_1 \parallel \ldots \parallel B_k)$, where $B_1 = C_1$ and $B_j = C_j \oplus f_{i,j-1}(B_{j-1})$ for each $2 \leq j \leq k$. It is easily seen that $\pi_{3,i}$ is not an involution.

For s-tuples of functions $(h_1, h_2, \ldots, h_s)$, define

$$\psi_3(h_s, \ldots, h_2, h_1) = g_{3,s} \circ \ldots \circ g_{3,2} \circ g_{3,1}$$

Note that ψ_3 is an invertible permutation on Σ^{kn}, and its inverse $\bar{\psi}_3$ is defined by

$$\bar{\psi}_3(h_1, h_2, \ldots, h_s) = \bar{g}_{3,1} \circ \bar{g}_{3,2} \circ \ldots \circ \bar{g}_{3,s}$$

5.4.2 Super-Pseudorandomness of Type-1 Transformations

Zheng, Matsumoto and Imai investigated the construction of provably secure block ciphers [Zheng *et al.*, 1990d]. A by-product of their work was the construction of super-pseudorandom permutation generators from type-2 and type-3 transformations, where they proved that $s \geq k+2$ rounds are required. In this subsection, we investigate the necessary and sufficient conditions for the construction of super-pseudorandom permutations based on type-1 Feistel permutations. These conditions will be presented after some preliminary observations. Moreover, we show that k^2 rounds of such permutations yield a super-pseudorandom permutation generator.

A Few Observations

It can be shown that $2k - 1$ rounds of type-1 transformations, where each round is associated with a randomly and independently chosen function from F_n, is a pseudorandom permutation.

The following lemma was proved by Zheng, Matsumoto, and Imai in [Zheng et al., 1990d] and formally represents the above statement.

Lemma 5.3 *Let Q be a polynomial in n and let C_{kn} be an oracle circuit with $Q(n) < 2^n$ oracle gates; then*

$$| \operatorname{Prob}\{C_{kn}[P_{kn}] = 1\} - \operatorname{Prob}\{C_{kn}[\psi_1(f_{2k-1}, \ldots, f_2, f_1)] = 1\} | \leq \frac{(k-1)Q^2(n)}{2^n}$$

where $f_1, f_2, \ldots, f_{2k-1} \in_r F_n$.

Although the above lemma states that $\psi_1(f_{2k-1}, \ldots, f_2, f_1)$ is pseudorandom, it is interesting to note that this structure is not super-pseudorandom. This is proved in the following lemma, where a super-distinguishing circuit is presented.

Lemma 5.4 *For any $f_{2k-1}, \ldots, f_2, f_1 \in_r F_n$, there is super-distinguishing circuit SC_{kn} for $\psi_1(f_{2k-1}, \ldots, f_2, f_1)$.*

Proof : Let $B_1, B_2, \ldots, B_k$ be strings of length n. The super-distinguishing circuit has two oracles, a normal oracle and an inverse oracle. The input to the normal oracle is $B_1 \parallel B_2 \parallel \ldots \parallel B_k$. Let $S_1 \parallel S_2 \parallel \ldots \parallel S_k$ be the output of this oracle. Let the input to the inverse oracle gate be $S_1 \oplus \alpha \parallel S_2 \parallel \ldots \parallel S_k$ where α is an arbitrary n-bit string. The output of SC_{2n} is **1** if and only if the last n bits of the output from the inverse oracle gate are equal to $B_k \oplus \alpha$. It can be verified that the output of SC_{2n} is always **1** when the normal and inverse oracle gates are computed using ψ_1 for the normal oracle gate and $\overline{\psi}_1$ for the inverse oracle gate. On the other hand, if the oracle gates are computed using a permutation randomly chosen from P_{kn}, the output of SC_{kn} is **1** with probability $\frac{1}{2^n}$ (see Figure 5.12). $\square$ It can be easily verified that, by using an inverse oracle together with a normal oracle, the effect of

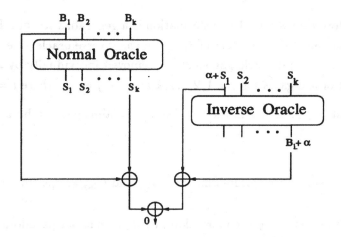

Figure 5.12: A Distinguishing Circuit for $\psi_1(f_{2k-1}, \ldots, f_2, f_1)$

f_{2k-1} is virtually removed. In other words, the super-distinguisher actually evaluates the inverse oracle with

$$\overline{\psi}_1(f_1, f_2, \ldots, f_{2k-2})$$

which is not pseudorandom by any means.

It can also be shown that the effect of $f_{2k-2}, \ldots, f_{k+2}$ and f_k can also be removed (individually) by procedures similar to those given in the proof of Theorem 5.1. If there existed a construction with type-1 transformations G_1, such that, after removing the last k random functions in G_1 and the first k random functions in $\overline{G}_1$, the remaining structures were pseudorandom, then G_1 would be a super-pseudorandom permutation.

Necessary and Sufficient Conditions

We now give the necessary and sufficient conditions for the super-pseudoran-domness of i rounds of type-1 transformations.

Theorem 5.3 *Let $G_1 = \psi_1(f_i, \ldots, f_1)$ be a pseudorandom permutation where $G_1 \in P_{kn}$ and consists of i rounds of type-1 transformations and $f_1, f_2, \ldots, f_i \in_r H_n$. Then G_1 is a super-pseudorandom permutation if and only if $G_{2,j} = \psi_1(f_i, \ldots, f_{j+1}) \circ L_{rot} \circ \psi_1(f_{j-1}, \ldots f_1)$ and $G_{3,j} = \overline{\psi}_1(f_1, \ldots, f_{i-j}) \circ R_{rot} \circ \overline{\psi}_1(f_{i-j+2}, \ldots, f_i)$ are pseudorandom permutations for $j = 1, 2, \ldots, k$ and $i - j \neq kl$, where $l = \lfloor \frac{i}{k} \rfloor$.*

Proof : Note that a type-1 transformation is a generalization of a DES-like permutation, and that the effect of f_{i-j+1} in the inverse oracle gates and the effect of f_j in normal oracle gates can be removed individually by applying normal and oracle gates for $j = 1, 2, \ldots, k$ and $i - j \neq kl$, where $l = \lfloor \frac{i}{k} \rfloor$.

To justify the theorem, the two following claims need to be proved for each j.

1. If G_1 is super-pseudorandom, then $G_{2,j}$ and $G_{3,j}$ are pseudorandom.

2. If $G_{2,j}$ and $G_{3,j}$ are pseudorandom, then G_1 is super-pseudorandom.

The validity of the above claims can be checked for each j. For instance, consider $j = 1$; then $G_{2,1} = \psi_1(f_i, \ldots, f_2) \circ L_{rot}$ and $G_{3,1} = \overline{\psi}_1(f_1, \ldots, f_{i-1}) \circ R_{rot}$. As in the proof of Lemma 5.1, it can be shown that $G_{2,1}$ and $G_{3,1}$ are independent if and only if they are pseudorandom. In addition, as in the proof of Theorem 5.1, it can be shown that, if $G_{2,1}$ and $G_{3,1}$ are independent, then G_1 is super-pseudorandom. For $j = 2, \ldots, k$ the proofs can be obtained by a similar method. Note that when $i - j = \lfloor \frac{i}{k} \rfloor k$, the effect of f_{i-j+1} cannot be removed with inverse and normal oracle gates, because of the structure of type-1 transformations; so it is not necessary to prove the above claims for it. Since all possible reductions of ψ_1 and $\overline{\psi}_1$ remain pseudorandom having even super-distinguishing circuits, then ψ_1 is a super-pseudorandom permutation. □

Although it was already stated in Lemma 5.3 that $2k-1$ rounds of type-1 permutations ψ_1 give a pseudorandom permutation, $3k - 2$ rounds of this transformation do not necessarily yield a super-pseudorandom permutation since $2k - 1$ rounds, or even $3k - 2$ rounds, of its inverse transformation $\overline{g}$ do not result in pseudorandomness. It can be shown that $k(k - 1) + 1$ rounds of $\overline{g}$ result in pseudorandomness. This is formally stated in the following lemma.

Lemma 5.5 *Let $\overline{\psi}_1$ be a permutation defined by*

$$\overline{\psi}_1(f_{k^2-k+1}, \ldots, f_2, f_1) = \overline{g}_{1,k^2-k+1} \circ \cdots \circ \overline{g}_{1,2} \circ \overline{g}_{1,1}$$

Let Q be a polynomial in n and let C_{kn} be an oracle circuit with $Q(n) < 2^n$ oracle gates; then

$$| \operatorname{Prob}\{C_{kn}[P_{kn}] = 1\} - \operatorname{Prob}\{C_{kn}[\bar{\psi}_1(f_{k^2-k+1}, \ldots, f_2, f_1)] = 1\} | \le$$
$$\le \frac{(k^2 - k + 1)Q^2(n)}{2^{n+1}}$$

where $f_1, f_2, \ldots, f_{k^2-k+1} \in_r H_n$.

Proof :   The proof of this lemma is very similar to the proof of Lemma 5.3 presented in [Zheng *et al.*, 1990d]. The method for proof is the same as the method developed in [Luby and Rackoff, 1988] for proving the pseudorandomness of $\psi(h, g, f)$.

Here, we use the notation from [Zheng, 1990]. Assume that C_{kn} is an oracle circuit with $Q(n)^2$ oracle gates, which are numbered $1, 2, \ldots, Q$. The inputs to the oracle gates are all different. Let Ω be the probability space on $(k^2 - k + 1)nQ$ bit strings with the uniform probability distribution. Any $\omega \in \Omega$ can be written as $\omega = \omega_1 \omega_2 \ldots \omega_{(k^2-k+1)nQ}$. For each $1 \le i \le Q$, $1 \le j \le 2k - 1$, define a random variable $X_{i,j}$ as follows:

$$X_{i,j}(\omega) = \omega_{b+1} \ldots \omega_{b+n}$$

where $b = jnQ + (i-1)n$. There is a total of $(k^2 - k + 1)Q$ such variables. For each $1 \le j \le k^2 - k + 1$, let $X_j(\omega) = X_{1,j}(\omega) \parallel X_{2,j}(\omega) \parallel \ldots \parallel X_{Q,j}(\omega)$. At a sample point $\omega \in \Omega$, P_i-gate is defined as follows:

P_i-gate:

The input is $(B_{i,1} \parallel B_{i,2} \parallel \ldots \parallel B_{i,k})$.

$u_{i,1} = min\{d : 1 \le d \le i \text{ and } B_{i,1} = B_{d,1}\}$.

We let $B_{i,2} = B_{i,1} \oplus X_{u_{i,1},1}$.

For $2 \le j \le k^2 - k + 1$, do the following operations.

If j is a multiple of k do:

$u_{i,j} = min\{d : 1 \le d \le i \text{ and } B_{i,k} = B_{d,k}\}$,

and let $B_{i,1} = B_{i,k} \oplus X_{u_{i,j},j}$

---

[2] $Q(n)$ is abbreviated to Q.

otherwise do:

$$u_{i,j} = min\{d : 1 \leq d \leq i \text{ and } B_{i,j \bmod k} = B_{d,j \bmod k}\},$$

and let $B_{i,j \bmod k+1} = B_{i,j \bmod k} \oplus X_{u_{i,j},j}.$

The output is $(B_{i,k} \parallel B_{i,1} \parallel \ldots \parallel B_{i,k-1}).$

Note that the structure of a P_i-gate is similar to $\overline{\psi}_1(f_{k^2-k+1}, \ldots, f_2, f_1).$ Let the random variable $C(\omega)$ be the output of the circuit C_{kn} when the oracle gates are evaluated by the above P_i-gates, and let $E(C)$ be the expectation of $C(\omega)$. Hence, the value of $E(C)$ is equal to the probability that $C(\omega) = 1.$

We now describe a random variable C' which is equal to the output bit of the distinguishing circuit C_{kn} when the oracle gates are evaluated with $\overline{\psi}_2(f_{k^2-k+1}, \ldots, f_2, f_1)$. Then we show that $E(C) = E(C').$

Let the random variable $C'(\omega)$ be the output of the circuit C_{kn} when the oracle gates are evaluated by the introduction of P_i'-gates. A P_i'-gate is described as follows:

P_i'-gate:

The input is $(B_{i,1} \parallel B_{i,2} \parallel \ldots \parallel B_{i,k}).$

$u_{i,1} = min\{d : 1 \leq d \leq i \text{ and } B_{i,1} = B_{d,1}\},$

and let $B_{i,2} = B_{i,1} \oplus X_{u_{i,1},1}.$

For $2 \leq j \leq k^2 - k + 1$, do the following operations.

    If j is a multiple of k do:

$$u_{i,j} = min\{d : 1 \leq d \leq i \text{ and } B_{i,k} = B_{d,k}\},$$

and let $X_{i,j}' = B_{i,k} \oplus X_{i,j},$

and let $B_{i,1} = B_{i,k} \oplus X_{u_{i,j},j}'$

    otherwise do:

$$u_{i,j} = min\{d : 1 \leq d \leq i \text{ and } B_{i,j \bmod k} = B_{d,j \bmod k}\},$$

and let $X_{i,j}' = B_{i,j \bmod k} \oplus X_{i,j},$

and let $B_{i,j \bmod k+1} = B_{i,j \bmod k} \oplus X_{u_{i,j},j}'$

The output is $(B_{i,k} \parallel B_{i,1} \parallel \ldots \parallel B_{i,k-1})$

It is clear that $E(C') = \text{Prob}\{C_{kn}[\overline{\psi}_2(f_{k^2-k+1}, \ldots, f_2, f_1)] = 1\}$. Now we show that $E(C) = E(C')$. Note that $X_{i,j}(\omega)$ has a uniform distribution on Σ^n. As, at each round, $B_{i,j}$ does not depend on $X_{i,j}$, then $X'_{i,j} = B_{i,j} \oplus X_{i,j}$ also has a uniform distribution on Σ^n. Hence, $E(C)$ and $E(C')$ are identical. Let A be the random variable which is defined to be the output of the distinguisher C_{kn} when the oracle gates are evaluated exactly the same way as in the definition of a P' gate, except that the output of the i-th oracle gate is $(X_{i,1} \parallel X_{i,2} \parallel \ldots \parallel X_{i,j})$. Because A is determined by C_{kn} when the output values from each oracle gate are independently and identically distributed random variables and because C_{kn} never repeats an input value to an oracle gate, $E(A) = \text{Prob}\{C_{kn}[H_{kn}] = 1\}$. Then, it follows that

$$\mid \text{Prob}\{C_{kn}[H_{kn}] = 1\} - \text{Prob}\{C_{kn}[\overline{\psi}_2(f_{k^2-k+1}, \ldots, f_2, f_1)] = 1\} \mid =$$
$$= \mid E(A) - E(C') \mid$$

For $\omega \in \Omega$, if there are pairs (d, i) with $1 \le d < i \le Q$ such that $B_{d,2} = B_{i,2}$, then X_1 is called *bad*. As there are Q oracle gates, then the probability that X_1 is bad is

$$\text{Prob}\{X_1(\omega) \text{ is bad }\} = \frac{Q(Q-1)}{2^{n+1}} \le \frac{Q^2}{2^{n+1}}$$

Similarly, for $\omega \in \Omega$, if there are pairs (d, i) with $1 \le d < i \le Q$ such that $B_{d,j+1} = B_{i,j+1}$, then X_j is called *bad*. The probability that X_j is bad is

$$\text{Prob}\{X_j(\omega) \text{ is bad }\} \le \frac{Q^2}{2^{n+1}}$$

If $X_j(\omega)$ is not bad for all $1 \le j \le k^2 - k + 1$, then $A(\omega) = C'(\omega)$. Thus we have

$$\mid \text{Prob}\{C_{kn}[H_{kn}] = 1\} - \text{Prob}\{C_{kn}[\overline{\psi}_2(f_{k^2-k+1}, \ldots, f_2, f_1)] = 1\} \mid \le$$
$$\le \frac{(k^2 - k + 1)Q^2}{2^{n+1}}$$

This completes the proof of the lemma.   □

Since, according to Theorem 5.3, by the application of normal and reverse oracle gates, the effect of at most $k - 1$ rounds of ψ and $k - 1$ rounds of $\overline{\psi}_1$ can be removed, $k^2 = k^2 - k + 1 + (k-1) > 2k - 1 + (k-1)$ rounds of type-1 transformations can resist super-distinguishing circuits. This is stated formally in the following theorem.

Theorem 5.4 *Let Q be a polynomial in n and SC_{kn} be a super-distinguishing circuit with $Q(n) < 2^n$ normal and inverse oracle gates; then*

$$| \operatorname{Prob}\{SC_{kn}[P_{kn}] = 1\} - \operatorname{Prob}\{SC_{kn}[\psi_1(f_{k^2}, \ldots, f_2, f_1)] = 1\} | < \frac{k^2 Q^2(n)}{2^n}$$

where $f_1, f_2, \ldots, f_{k^2} \in_r H_n$.

Proof :   To obtain a pseudorandom permutation, $G_{2,j}$ and $G_{3,j}$ should be pseudorandom for $j = 1, 2, \ldots, k-1$, where, for $\psi_1(f_{k^2}, \ldots, f_2, f_1)$,

$$G_{2,j} = \psi_1(f_{k^2}, \ldots, f_{j+1}) \circ L_{rot} \circ \psi_1(f_{j-1}, \ldots, f_1)$$

and

$$G_{3,j} = \overline{\psi}_1(f_1, \ldots, f_{k^2-j}) \circ R_{rot} \circ \overline{\psi}_1(f_{k^2-j+2}, \ldots, f_{k^2})$$

$G_{2,j}$ is partitioned into two parts where a part always consists of more than $2k - 1$ rounds. So, in the normal oracles, even if the effect of the other $k - 1$ rounds of ψ_1 could be removed, the remaining oracle gates would maintain pseudorandomness. $G_{3,j}$ is also partitioned into two parts where a part always consists of at least $k^2 - k + 1$ rounds; so, in the inverse oracles, even if the effect of the other $k - 1$ rounds of $\overline{\psi}_1$ could be removed, the remaining would maintain pseudorandomness. Hence, $G_{2,j}$ and $G_{3,j}$ are pseudorandom for all $j = 1, 2, \ldots, k-1$, and G_1 is a super-pseudorandom permutation. The probability that a super-distinguishing circuit outputs **1**, in the worst case, is equal to the probability that a distinguishing circuit for ψ_1 outputs **1**, plus the probability that a distinguishing circuit for $\overline{\psi}_1$ outputs **1**, and is less than $\frac{k^2 Q^2(n)}{2^n}$. $\square$

5.5   Conclusions and Open Problems

In the first part of this chapter, we presented the necessary and sufficient conditions for the construction of super-pseudorandom permutation generators based on DES-like permutations. If a block cryptosystem is super-pseudorandom, it is secure against the chosen plaintext/ciphertext attack which is a much stronger attack than a chosen plaintext attack. We also showed that $\psi(g, f, f)$, a cryptosystem which consists of DES-like permutations and is secure against chosen plaintext attacks, can be enhanced to a super-pseudorandom cryptosystem, that is, $\psi(g, g, f, f)$, by adding one more round of DES-like permutations. It still remains to be shown *how to construct a super-pseudorandom permutation from a single pseudorandom function*. We give a solution to this question in Chapter 6.

In the second part, we investigated the conditions for the super-pseudo-randomness of constructions based on type-1 generalized Feistel permutations. We showed that the composition of k^2 rounds of such permutations with k^2 pseudorandom function generators, yields a super-pseudorandom permutation generator. It can be shown that $\psi_1(f_{k-1}, \ldots, f_1, f_k, f_{k-1}, \ldots, f_1)$ is not pseudorandom although it consists of $2k - 1$ rounds of DES-like permutations. On the other hand, it can be conjectured that $\psi_1(f_2, \ldots, f_2, f_1, \ldots, f_1)$, where f_1 is used in k rounds and f_2 is used in $k - 1$, is pseudorandom. It remains to be discovered *what is the minimum number of random functions needed to achieve super-pseudorandomness with k^2 rounds of type-1 transformations.*

Chapter 6

A Sound Structure

6.1 Introduction

As we mentioned in Chapter 4, Luby and Rackoff employed a structure with three rounds of DES-like permutations to build a pseudorandom permutation generator, and considered their result a justification for the application of DES-like permutations in the design of DES. They also proved that four rounds of such permutations would provide a super-pseudorandom permutation generator. An implication of this result, is that a greater number of rounds gives better security. In this chapter, we show how to construct a super-pseudorandom permutation generator from a single pseudorandom function generator. This structure is obtained by some modifications in the structure proposed by Luby and Rackoff.

Clearly, the composition of two or more Luby and Rackoff permutation generators is also pseudorandom. One would expect that for the resulting structure, the probability of distinguishing drops to zero if a large enough number of Luby-Rackoff generators is used. Although, the probability of distinguishing can be made as small as requested, it will never drop to zero. Since for any n there is a finite number of compositions after which the alternating group $A_{2n} \subset P_{2n}$ would be generated, there should be a better way to design a permutation generator. Pieprzyk and Sadeghiyan constructed such an improved version of the Luby and Rackoff construction [Pieprzyk and Sadeghiyan, 1991]. In Section 6.2, some properties of the Luby and Rackoff construction together with a brief explanation of their proof are examined. Then the improved construction of Pieprzyk and Sadeghiyan is presented, and it is shown that the composition of two Luby and Rackoff structures with four random function generators and two random permutation generators provides a perfect randomizer. At the end, we present the construction of the super-pseudorandom permutation generator with a single pseudorandom function generator. We recommend that this structure be used in the design of block ciphers, as it exhibits better cryptographic strength and has a simple configuration.

The results of this chapter have been presented in [Pieprzyk and Sadeghiyan, 1991] and [Sadeghiyan and Pieprzyk, 1992].

6.2   Preliminaries

In Chapter 4, we defined DES-like permutations and their compositions, where, given a sequence of functions $f_1, f_2, \cdots, f_i \in H_n$, the composition of their DES-like permutations ψ is defined as

$$\psi(f_i, f_{i-1}, \ldots, f_1) = D_{2n,f_i} \circ D_{2n,f_{i-1}} \circ \cdots \circ D_{2n,f_1}$$

with $\psi(f_i, f_{i-1}, \ldots, f_1) \in P_{2n}$. For the selection of functions f_j for $j = 1, \cdots, i$, there are two possible cases.

- The functions are chosen randomly from different pseudorandom function generators, that is $f_i \in_r F_n$. For conciseness, we will say that the functions are pseudorandom. The resulting permutation ψ is pseudorandom for $i = 3, 4, \cdots$.

- The functions are chosen randomly from the set of all functions in n, that is, $f_i \in_r H_n$. For conciseness, we will say the functions are random. The resulting permutation generator is called a *randomizer*.

To draw some conclusions about the quality of the structured permutations based on either pseudorandom or truly random functions, distinguishing circuits introduced in Chapter 4 are used. Structured permutation generators can first be assessed applying truly random functions. Then, if the structure is sound, pseudorandom functions may replace the truly random ones. For example, Luby and Rackoff first considered the permutation generator $\psi(f, g, h)$ with three rounds of DES and three random functions f, g, h. They proved that

$$| \text{Prob}\{C_{2n}[P_{2n}] = 1\} - \text{Prob}\{C_{2n}[\psi(h, g, f)] = 1\} | \leq \frac{m^2}{2^n} \qquad (6.1)$$

where m is the number of oracle gates and $m \leq 2^n$. When $f, g, h \in_r H_n$, this structure is called an L-R randomizer. They showed that the structure $\psi(f, g, h)$ can be "transparent" to the input and proved the necessary conditions for "a leakage" input information to the output. To prove this, they chose an $\omega \in \Omega$, where $\Omega = \Sigma^{3mn}$ is the sample space with uniform probability distribution, that is, for all $\omega \in \Omega$, $\text{Prob}\{\omega\} = \frac{1}{2^{3mn}}$. Then ω was divided into three $(m \times n)$-bit strings, called X, Y, Z, respectively, and each was divided into m n-bit segments. The strings X, Y and Z were applied to construct gates with a similar structure to $\psi(h, g, f)$. For there to be a leakage of information from the input to the output of the gate, two of the m segments of X or Y must have the same value. This is called a collision of those two segments. If there is no leakage of the input to the output, they said that ω *is preserving*, where any distinguishing circuit cannot make a decision whether the generator used to evaluate the oracle gates was P_{2n} or $\psi(h, g, f)$. Then they showed that if the random functions were replaced by pseudorandom ones, then the probability of distinguishing between outputs would remain less than 1 over any polynomial in n.

ω *is preserving* means that all the outputs of the oracle gates are independent from the input (and from each other). When ω *is NOT preserving* means that there is at least one pair of oracle gates such that their outputs are related to their inputs and this relation can be used to distinguish ψ from P_{2n}.

Note that if ω *is preserving* in a distinguishing circuit, then the distinguisher cannot find any pair of oracle gates with outputs related to its inputs. Luby and Rackoff showed that the leakage of the input happens only in two cases, Y *is bad* or X *is bad*.

Y *is bad* if there is a pair of oracle gates (O_i, O_j) such that the random function g collides, which is the case where g gives the same output for two different inputs. Figure 6.1 depicts this case. It happens when the input random variables are such that $R = R_i = R_j$, but $L_i \neq L_j$. It is obvious that

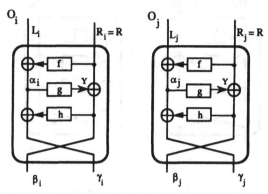

Figure 6.1: Y is Bad in Two Oracle Gates (O_i, O_j)

$$\alpha_i = L_i \oplus f(R) \quad \text{and} \quad \alpha_j = L_j \oplus f(R)$$

and, as $L_i \neq L_j$, then $\alpha_i \neq \alpha_j$. Hence the random function g assigns two independent random variables $g(\alpha_i)$ and $g(\alpha_j)$, and the outputs β_i, β_j are independent from the input. The outputs γ_i, γ_j are also independent only if the random variables $g(\alpha_i)$, $g(\alpha_j)$ take on different values. Otherwise, if $\beta_i = \beta_j$ (this may happen with probability $\frac{1}{2^n}$ for a single pair of oracle gates), γ_i, γ_j are related. This may happen only if the function g collides, that is,

$$g(\alpha_i) = g(\alpha_j) = Y$$

and then

$$\gamma_i = \alpha_i \oplus h(Y)$$
$$\gamma_j = \alpha_j \oplus h(Y)$$

In this case, $\gamma_i \oplus \gamma_j = L_i \oplus L_j$ always.

The second possibility for input information leakage to the output happens when X *is bad* (see Figure 6.2). It can happen only if $R_i \neq R_j$. The random function f assigns two independent random variables $f(R_i)$ and $f(R_j)$; as a result, the output variables γ_i, γ_j are independent of the input. The input information can pass through β to the output if $\alpha = \alpha_i = \alpha_j$ (this happens with probability $\frac{1}{2^n}$ for a single pair of gates). Then

$$\beta_i = R_i \oplus g(\alpha)$$
$$\beta_j = R_j \oplus g(\alpha)$$

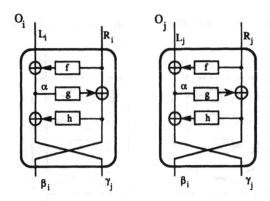

Figure 6.2: X is Bad in Two Oracle Gates (O_i, O_j)

In this case, $\beta_i \oplus \beta_j = R_i \oplus R_j$ always.

The distinguisher can apply two strategies: the first one *"hunting for (Y is bad)"* or the second one *"hunting for (X is bad)"*. Luby and Rackoff calculated that the probability P_Y that Y *is bad* in at least one pair of gates is

$$P_Y \leq \frac{m(m-1)}{2} \frac{1}{2^n}$$

where $\frac{m(m-1)}{2}$ is the number of different pairs of gates if the distinguisher has m oracle gates. In the second strategy, the distinguisher selects different R_i for all oracle gates and the probability P_X that X *is bad* in at least one pair of oracle gates is

$$P_X \leq \frac{m(m-1)}{2} \frac{1}{2^n}$$

Obviously if a distinguisher applies some mixed strategy, then

$$\text{Prob}[\omega \text{ is NOT preserving}] \leq P_Y + P_X \leq \frac{m^2}{2^n}$$

Now consider a randomizer $\Psi_2 = \psi_1(f, g, h) \circ \psi_2(f, g, h)$ which is constructed from two L-R randomizers. ω *is NOT preserving* in Ψ_2 if there is at least one pair of oracle gates O_i, O_j for which Y *is bad* or X *is bad* in the first randomizer ψ_1. Figure 6.3 shows the pair with Y *is bad* (note that $R_i = R_j = R$). Clearly the outputs $\gamma_{i_2}, \gamma_{j_2}$ are independent of the input. β_{i_2} and β_{j_2}. However, the outputs may be related if $\beta_{i_1} = \beta_{j_1}$ (with probability $\frac{1}{2^n}$ for a single pair of oracles (O_i, O_j)) and $\alpha_{i_2} = \alpha_{j_2}$ (this happens with probability $\frac{1}{2^n}$ in a single pair of oracle gates). Therefore the probability of

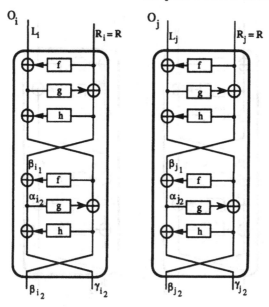

Figure 6.3: Y is Bad in Two Oracle Gates (O_i, O_j) for Ψ_2

Y being bad in a single pair of oracle gates is $\frac{1}{2^{2n}}$. Considering all the possible pairs of gates, we can conclude that the probability P_Y that Y *is bad* in Ψ_2 is

$$P_Y = \frac{m(m-1)}{2} \frac{1}{2^{2n}}$$

The second case when X *is bad* in ψ_1 is presented in Figure 6.4, where $R_i \neq R_j$. Clearly, the outputs β_{i_2}, β_{j_2} are independent from the input. If X *is bad* in (O_i, O_j), then $\alpha_{i_1} = \alpha_{j_1} = \alpha$ and consequently $\beta_{i_1} = R_i \oplus g(\alpha)$ and $\beta_{j_1} = R_j \oplus g(\alpha)$. The function h in ψ_1 generates two independent random variables. Thus $\gamma = \gamma_{i_1} = \gamma_{j_1}$ with probability $\frac{1}{2^n}$ and the relation to the input remains. The random function g in ψ_2 assigns two independent random variables and the outputs $\gamma_{i_2}, \gamma_{j_2}$ are related only if $\beta = \beta_{i_2} = \beta_{j_2}$ (with probability $\frac{1}{2^n}$). Therefore X *is bad* in Ψ_2 for a single pair of oracle gates with probability $\frac{1}{2^{3n}}$. If a distinguisher uses some strategy to tell apart the tested permutation generator, the probability of its success is

$$\text{Prob}[\omega \text{ is NOT preserving in } \Psi_2] \leq \frac{m(m-1)}{2} \left(\frac{1}{2^{2n}} + \frac{1}{2^{3n}} \right)$$

Hence, we have proved the following theorem.

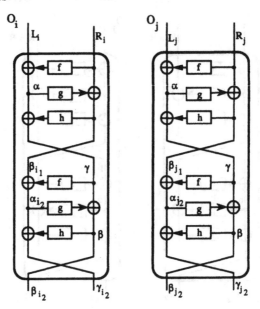

Figure 6.4: X is Bad in Two Oracle Gates (O_i, O_j) for Ψ_2

Theorem 6.1 *The randomizer $\Psi_2 = \psi_1(f, g, h) \circ \psi_2(f, g, h)$, where $f, g, h \in_r H_n$, does not have a distinguisher and*

$$| \operatorname{Prob}\{C_{2n}[P_{2n}] = 1\} - \operatorname{Prob}\{C_{2n}[\Psi_2] = 1\} | \leq \frac{m^2}{2} \left(\frac{1}{2^{2n}} + \frac{1}{2^{3n}} \right) \quad (6.2)$$

where $m \leq 2^n$ is the number of oracle gates in the distinguisher.

It is easy to generalize the previous theorem for the composition of $k = 2, 3, 4, \cdots$ L-R randomizers. As the parameter k grows, the probability of distinguishing becomes smaller for the generator

$$\Psi_k = \underbrace{\psi_1(f_1, g_1, h_1) \circ \ldots \circ \psi_k(f_k, g_k, h_k)}_{k} \ .$$

Theorem 6.2 *The randomizer Ψ_k, where $f_i, g_i, h_i \in_r H_n$ and $i = 1, \ldots, k$, does not have a distinguisher and*

$$| \operatorname{Prob}\{C_{2n}[P_{2n}] = 1\} - \operatorname{Prob}\{C_{2n}[\Psi_k] = 1\} | \leq \frac{m^2}{2} \left(\frac{1}{2^{kn}} + \frac{1}{2^{(2k-1)n}} \right) \quad (6.3)$$

where $m \leq 2^n$ is the number of oracle gates in the distinguisher.

6.3   Perfect Randomizers

In Section 6.2, we saw that the composition of L-R randomizers does not
have any distinguisher with m oracle gates, where $m \leq 2^n$. There is always
a small probability of success no matter how many elementary randomizers
are used. Since for any parameter n there is a finite number of compositions,
after which the alternating group $A_{2n} \subset P_{2n}$ is generated [1], there must be a
better way to design a permutation generator.

In this section, we show how to improve the L-R randomizer to obtain
the so-called *perfect randomizer*. Perfectness is defined as follows [Pieprzyk
and Sadeghiyan, 1991]:

Definition 6.1 *A randomizer is perfect if, for all oracle gates used by the
distinguisher, their outputs are independent of their inputs and independent
of each other.*

In the next theorem, a modification in the L-R randomizer structure is made
and it is shown that the change does not diminish its quality. At the same
time the modified randomizer has a further desired property, as one output
branch (γ output) is always independent of the input.

Theorem 6.3 *Let $f, h \in_r H_n$ and $g^* \in_r P_n$; then the randomizer $\psi(f, g^*, h)$,
does not have any distinguisher and*

$$| Pr[C_{2n}(P_{2n})] - Pr[C_{2n}(\psi)] | \leq \frac{m^2}{2^n} \qquad (6.4)$$

where $m \leq 2^n$ is the number of oracle gates in the distinguisher.

Proof :   For the proof of the above theorem, the main idea that Luby and
Rackoff used in their proof is adopted in the following explanations. Using
their notation, it can be said that ω *is NOT preserving* in ψ if Y *is bad* or
X *is bad*. If Y *is bad* in a pair of oracle gates (O_i, O_j), then $R = R_i = R_j$
and $\alpha_i \neq \alpha_j$ (see Figure 6.5). Therefore the random permutation g^* assigns

---

[1]The number of compositions may be exponential in n. In [Pieprzyk and Zhang, 1990],
it is shown that it is possible to produce $(2^n)!$ different permutations, having $(2^{\frac{n}{2}})!$ generators.

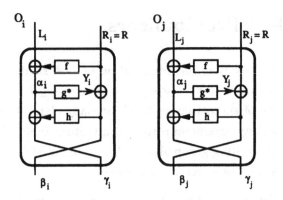

Figure 6.5: Y is Bad in Two Oracle Gates (O_i, O_j) for $\psi(f, g^*, h)$

two different random variables Y_i, Y_j, which never collide, and the random function h generates two independent random variables, that is, the outputs γ_i, γ_j are independent of the input. The distinguisher, however, can work on β_i, β_j as they are generated according to a different probability distribution (they are random permutations). Clearly, β_i and β_j are always different if the oracle gates are evaluated by ψ. If all the oracle gates are evaluated by P_{2n}, then β_i may collide, where $i = 1, 2, \ldots, m$. When oracle gates are evaluated by P_{2n}, the probability that β_i do not collide is

$$\frac{2^n!}{2^{nm}(2^n - m)!}$$

Thus the probability P_Y, that is, Y is bad and the distinguisher succeeds in finding a collision when the oracle gates are evaluated by F_{2n}, is

$$P_Y = 1 - \frac{2^n!}{2^{nm}(2^n - m)!} \leq \frac{m(m+1)}{2^{n+1}}$$

Consider the second case when X is bad, where $R_i \neq R_j$. This case is identical to that in Figure 6.2. The random function f assigns two independent random variables $f(R_i)$ and $f(R_j)$. The outputs γ_i and γ_j are independent of the input. The probability P_X is precisely the same as for the original L-R randomizer. Therefore

$$\text{Prob}[\omega \text{ is NOT preserving in } \Psi_2] \leq P_X + P_Y \leq \frac{m(m-1)}{2} \frac{1}{2^n} + \frac{m(m+1)}{2^{n+1}}$$

and the final result follows.   □

Now we are ready for the main theorem of this section.

Theorem 6.4 *Let $f_1, f_2, h_1, h_2 \in_r H_n$ and $g_1^*, g_2^* \in_r P_n$. The randomizer $\Psi_2^* = \psi_1(f_1, g_1^*, h_1) \circ \psi_2(f_2, g_2^*, h_2)$ is perfect when the number of oracle gates $m \leq 2^n$.*

Proof :   If all $\frac{m(m-1)}{2}$ possible pairs of oracle gates are considered and none of the pairs is transparent to the input, it means that Ψ_2 is perfect, or, in Luby and Rackoff's terms, ω *is preserving*. Consider a single pair of

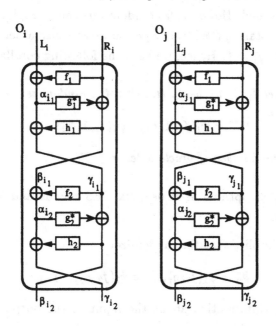

Figure 6.6: Two Oracle Gates (O_i, O_j) Evaluated by Ψ_2^*

oracle gates (O_i, O_j). We are going to show that its outputs $(\beta_{i_2}, \beta_{j_2})$ and $(\gamma_{i_2}, \gamma_{j_2})$ are independent of the input variables (see Figure 6.6). According to the previous lemma, the outputs γ_{i_1} and γ_{j_1} are always independent of the input; so are the outputs $(\beta_{i_2}, \beta_{j_2})$.

Now take the randomizer ψ_2 which is fed by two pairs $\beta_{i_1}, \gamma_{i_1}$ and $\beta_{j_1}, \gamma_{j_1}$. There are two possible cases.

1. γ_{i_1} and γ_{j_1} assume different values (this happens with probability $1 - \frac{1}{2^n}$). It turns out that f_2 assigns two independent random variables, and γ_{i_2} and γ_{j_2} are independent of the input.

2. γ_{i_1} and γ_{j_1} have the same value (this happens with probability $\frac{1}{2^n}$). Thus $\alpha_{i_2} \neq \alpha_{j_2}$ and the permutation g^* generates two different values. Finally, the random function h_2 makes γ_{i_2} and γ_{j_2} independent of the input.

$\square$

Random permutations g_1^*, g_2^* play an important role as far as single randomizers are concerned. However, Ψ_2 randomizers can be replaced by a fixed permutation, for instance, the identity permutation. Such a permutation can be written as $g_1^* = g_2^* = 1$. Hence, we have the following corollary.

Corollary 6.1 *Let $f_1, f_2, h_1, h_2 \in_r H_n$; then the randomizer*

$$\Psi_2^{**} = \psi_1(f_1, 1, h_1) \circ \psi_2(f_2, 1, h_2)$$

is perfect, when the number of oracle gates $m \leq 2^n$.

The structure Ψ_2^{**} is optimal as it uses six DES rounds and four different random functions.

Perfectness implies that the randomizer

$$\Psi_2^* = \psi_1(f_1, g_1^*, h_1) \circ \psi_2(f_2, g_2^*, h_2)$$

does not leak any information about the input to the output. Luby and Rackoff said that, in this case, ω *is preserving*. When ω *is preserving*, the distinguisher works only on the knowledge of the output (it obviously selects different input values but their values are not important).

From the above work, [Pieprzyk and Sadeghiyan, 1991] draw six conclusions about the design of pseudorandom permutation generators. We draw attention to three of these conclusions which can help in the design of a super-pseudorandom permutation generator from a single pseudorandom function generator, or a sound structure to be applied in the design of block-cipher-based hash schemes.

1. As the outputs of all oracle gates (evaluated by Ψ_2^* or $\Psi^*{}_2$) are independent random variables, the knowledge of their inputs does not provide any useful information to the distinguisher.

2. Most of the DES-type cryptosystems use the structure $\psi(f_1, f_2, \cdots, f_k)$, where the f_i $(i = 1, \cdots, k)$ are functions generated by a short cryptographic key. The functions are neither random nor pseudorandom. As far as random functions are employed, the structure

$$\psi(f_1, 1, f_2, f_3, 1, \cdots f_{k-1}, 1, f_k)$$

is better than $\psi(f_1, f_2, \cdots, f_k)$, as it is a perfect randomizer.

3. The alternating group A_{2n} of the group of all permutations P_{2n} can be generated using a finite number of concatenations of ψ [Pieprzyk and Zhang, 1990]. Thus

$$\mid \text{Prob}[C_{2n}(P_{2n})] - \text{Prob}[C_{2n}(\Psi_2^*)] \mid = 0$$

Note that $\mid Pr[C_{2n}(F_{2n})] - Pr[C_{2n}(\Psi_2^*)] \mid > 0$, as it is possible to design a distinguisher which can tell apart F_{2n} from Ψ_2^* with small probability. The distinguisher tries to get the same output in two different oracle gates for two different messages. It can succeed only if the oracle gates are evaluated by F_{2n} (for oracle gates evaluated by Ψ_2^*, any output is different for different input).

So far we have presented the result of Pieprzyk and Sadeghiyan for the construction of a perfect randomizer. We included this result and related explanations in order to use the recommended structure for the construction of a super-pseudorandom permutation generator from a single pseudorandom function generator based on DES-like permutations.

6.4   A Construction for Super-Pseudorandom Permutation Generators

In Chapter 5, the necessary and sufficient conditions for the construction of super-pseudorandom permutation generators based on DES-like permutations were investigated. It was also shown that $\psi(g, g, f, f)$ is super-pseudorandom, it is a structure with two pseudorandom functions and four DES-like permutations. This result was an improvement upon the result shown by Luby and Rackoff, where they demonstrated that $\psi(h, g, f, e)$ with four pseudorandom function generators is a super-pseudorandom permutation generator. However the question of how to construct super-pseudorandom permutations from a single pseudorandom function remained an open problem.

In the remainder of this chapter, we answer the above question and present a construction based on a single pseudorandom function which is super-pseudorandom. We take advantage of a structure $\psi(h, 1, f, h, 1, f)$ which is similar to that of the perfect randomizer presented earlier in Section 6.3, where not only the output is independent of the input but also the two branches of the output are independent of each other. First we show that the above structure provides us with a super-pseudorandom permutation generator. Then, we present a construction based on a single pseudorandom function, which replaces one of the pseudorandom functions with a two-fold composition of the other one, that is, $\psi(f^2, 1, f, f^2, 1, f)$, which is indistinguishable from the previous one. Finally, we show that the construction is super-pseudorandom. Hence, it is possible to construct a super-pseudorandom permutation from a single pseudorandom function, where we need six rounds of DES-like permutations and six references to the pseudorandom function.

6.4.1 Super-Pseudorandomness of $\psi(h, 1, f, h, 1, f)$

To construct a super-pseudorandom permutation generator based on a single pseudorandom function, we first show that $G_1 = \psi(h, 1, f) \circ \psi(h, 1, f)$ is a super-pseudorandom permutation generator. Then we show that, if f^2 is substituted for h, $\mathcal{G}_1 = \psi(f^2, 1, f, f^2, 1, f)$ is also super-pseudorandom. To show that G_1 is a super-pseudorandom permutation generator, we first show that $G_2 = \psi(h, 1, f, h, 1)$ and $G_3 = \psi(f, 1, h, f, 1)$ are not only pseudorandom but also independent permutations. Then we show that G_1 is super-pseudorandom.

Lemma 6.1 *Let $h, f \in_r H_n$ be independent random functions and $G_2 = \psi(h, 1, f, h, 1)$. Then*

$$| \operatorname{Prob}\{C_{2n}[G_2] = 1\} - \operatorname{Prob}\{C_{2n}[P_{2n}] = 1\} | \leq \frac{m^2}{2^n} + \frac{m^2}{2^{2n}}$$

where C_{2n} is any polynomial size distinguishing circuit with $m < 2^n$ oracle gates.

Proof : When the distinguisher examines an oracle, the input is a $2n$ bit string $(L \parallel R)$ and the output is a $2n$ bit string $(S \parallel T)$ where

$$S = L \oplus R \oplus f(L \oplus h(L \oplus R)) \oplus h(R \oplus h(L \oplus R) \oplus f(L \oplus h(L \oplus R)))$$
$$T = R \oplus h(L \oplus R) \oplus f(L \oplus h(L \oplus R))$$

For two different experiments, $(L_i \parallel R_i)$ should be different from $(L_j \parallel R_j)$; so either $L_i \neq L_j$ or $R_i \neq R_j$ or both are different. If there is no leakage of information from the input to the output, the distinguisher cannot distinguish the generator used to evaluate the oracle gates from a random generator. Leakage of information happens when there is at least one pair of oracle gates such that their outputs are related to their inputs. Let X be a random variable denoting the ouput of h, the random function in the second round of the DES-like structure of G_2, and let Y be a random variable denoting the output of f, the random function in the third round of DES-like structure of G_2 (see Figure 6.7). Leakage of information happens in two cases.

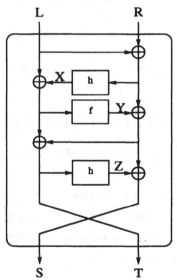

Figure 6.7: Random Variables X and Y in G_2

1. X **is bad.** This happens when there is a pair of oracle gates O_i, O_j with the input random variables $R_i \neq R_j$ and $L_i = L_j = L$ such that

the random function h collides. Hence, for this case, we assume that $x_i = x_j = x$ where $x_i = h(L \oplus R_i)$ and $x_j = h(L \oplus R_j)$. It is obvious that, if X is bad, then $T_i \oplus T_j = R_i \oplus R_j$ always. The probability that h collides in a pair of oracle gates among m oracles is equal to $\frac{m(m-1)}{2} \frac{1}{2^n}$.

2. Y is bad. This happens when there is a pair of oracle gates O_i, O_j with the input random variables $R_i = R_j = R$ and $L_i \neq L_j$ such that the random function f collides. Hence, for this case, we assume that $y_i = y_j = y$ where $y_i = f(L_i \oplus h(L_i \oplus R))$ and $y_j = f(L_j \oplus h(L_j \oplus R))$. It is obvious that, if Y is bad and X is also bad, then $T_i \oplus T_j = L_i \oplus L_j$ always. The probability that f collides and also h collides in a pair of oracle gates among m oracles is equal to $\frac{m(m-1)}{2} \frac{1}{2^{2n}}$.

The probability that a distinguishing circuit for G_2 can be constructed is equal to $\frac{m(m-1)}{2^{n+1}} + \frac{m(m-1)}{2^{2n+1}}$. On the other hand, when a permutation p is chosen randomly, the probability that p can satisfy the distinguishing circuit relation is $\frac{m}{2^n} + \frac{m}{2^{2n}}$. So, an upper bound on the probability of distinguishing is

$$| \operatorname{Prob}\{C_{2n}[P_{2n}] = 1\} - \operatorname{Prob}\{C_{2n}[\psi(h, 1, f, h, 1)] = 1\} | \leq \frac{m^2}{2^n} + \frac{m^2}{2^{2n}}$$

$\square$

Lemma 6.2 Let $h, f \in_r H_n$ be independent random functions and $G_3 = \psi(f, 1, h, f, 1)$. Then

$$| \operatorname{Prob}\{C_{2n}[G_3] = 1\} - \operatorname{Prob}\{C_{2n}[P_{2n}] = 1\} | \leq \frac{m^2}{2^n} + \frac{m^2}{2^{2n}}$$

where C_{2n} is a distinguishing circuit with $m < 2^n$ oracle gates.

Proof : Since the structure of G_3 is exactly the same as the structure of G_2 except that the roles of h and f are reversed, a proof similar to that of Lemma 6.1 can be given for the probability of distinguishing of G_3 from a random permutation, and is omitted here. $\square$

Note that the above lemma is an instance of Lemma 6.1, and when m is a polynomial in n, the probability of distinguishing G_2 or G_3 from a random permutation becomes less than $\frac{1}{n^{c_2}}$ for any constant c_2, and sufficiently large n.

Lemma 6.3 *Let $h, f \in_r H_n$ be independent random functions and $G_2 = \psi(h, 1, f, h, 1)$ and $G_3 = \psi(f, 1, h, f, 1)$; then*

$$| \operatorname{Prob}\{DC_{2n}[G_2, G_3] = 1\} - \operatorname{Prob}\{DC_{2n}[P_{2n}, P_{2n}] = 1\} | \leq \frac{m^2}{2^{2n}}$$

where DC_{2n} is a D-distinguishing circuit with two types of oracle gates and $m < 2^n$ the number of oracle gates.

Proof : As the distinguisher has two types of oracle gates, one probability is calculated when two permutations are chosen independently and randomly from P_{2n} and are used to evaluate the oracle gates. The other probability is calculated when the distinguisher chooses f and h independently and randomly from H_n and uses them in the G_2 and G_3 structures, which are applied for the evaluation of the oracle gates. When the distinguisher examines an oracle, the input is a $2n$ bit string $(L \parallel R)$ and the output is a $2n$ bit string $(S \parallel T)$.

When G_2 is examined, the output is

$$
\begin{aligned}
S &= L \oplus R \oplus f(L \oplus h(L \oplus R)) \oplus h(R \oplus h(L \oplus R) \oplus f(L \oplus h(L \oplus R))) \\
T &= R \oplus h(L \oplus R) \oplus f(L \oplus h(L \oplus R))
\end{aligned}
$$

and when G_3 is examined, the output is

$$
\begin{aligned}
S &= L \oplus R \oplus h(L \oplus f(L \oplus R)) \oplus f(R \oplus f(L \oplus R) \oplus h(L \oplus f(L \oplus R))) \\
T &= R \oplus f(L \oplus R) \oplus h(L \oplus f(L \oplus R))
\end{aligned}
$$

As both G_2 and G_3 are pseudorandom, there is no distinguishing circuit with one type of oracle gate for G_2 or for G_3. The D-distinguisher could only make a decision if there were at least one pair of oracle gates whose outputs were related to each other.

Let X_2 be a random variable denoting the ouput of h, the random function in the second round of the DES-like structure of G_2, and X_3 be a random variable denoting the output of f, the random function in the second round of the DES-like structure of G_3. Let Y_2 be a random variable denoting the output of f, the random function in the third round of DES-like structure of G_2, and Y_3 be a random variable notating the output of h, the random function in the third round of DES-like structure of G_3 (see Figure 6.8). The distinguisher can make a decision in either of the two following cases.

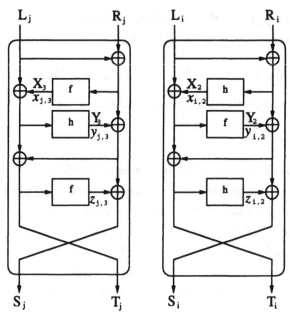

Figure 6.8: The Random Variables X_3, Y_3 in G_3 and X_2, Y_2 in G_2

1. When there is a pair of oracle gates O_i, O_j such that the random functions f and h collide. For this case we assume that $x_{i,2} = x_{j,3} = x$ where $x_{i,2} = h(L_i \oplus R_i)$ and $x_{j,3} = f(L_j \oplus R_j)$ and $y_{i,2} = y_{j,3} = y$ where $y_{i,2} = f(L_i \oplus h(L_i \oplus R_i))$ and $y_{j,3} = h(L_j \oplus f(L_j \oplus R_j))$. In this case, when the input random variables $R_i \neq R_j$ and $L_i = L_j = L$, then $T_i \oplus T_j = R_i \oplus R_j$ always, and when the input random variables $R_i = R_j = R$ and $L_i \neq L_j$ then $S_i \oplus S_j = L_i \oplus L_j$ always. The probability that f and h collide in a pair of oracle gates among m oracles is equal to $\frac{m(m-1)}{2} \frac{1}{2^{2n}}$.

2. When there is a pair of oracle gates O_i, O_j such that the random function f collides and the random function h collides. For this case we assume that $x_{i,2} = y_{j,3}$, where $x_{i,2} = h(L_i \oplus R_i)$ and $y_{j,3} = h(L_j \oplus f(L_j \oplus R_j))$, and $y_{i,2} = x_{j,3}$, where $y_{i,2} = f(L_i \oplus h(L_i \oplus R_i))$ and $x_{j,3} = f(L_j \oplus R_j)$ In this case, when the input random variables $R_i \neq R_j$ and $L_i = L_j = L$, then $T_i \oplus T_j = R_i \oplus R_j$ always. The probability that f collides and h also collides in a pair of oracle gates among m oracles is equal to $\frac{m(m-1)}{2} \frac{1}{2^{2n}}$.

The probability that a D-distinguishing circuit could be constructed for (G_2, G_3) would be equal to $\frac{m(m-1)}{2^{2n+1}} + \frac{m(m-1)}{2^{2n+1}}$. On the other hand, when two permutations p_1 and p_2 are chosen independently and randomly, the probability that they can satisfy the distinguishing circuit relation is $\frac{2m}{2^{2n}}$. So, an upper bound on the probability of distinguishing is

$$| \text{Prob}\{DC_{2n}[P_{2n}, P_{2n}] = 1\} - \text{Prob}\{DC_{2n}[G_2, G_3] = 1\} | \leq \frac{m^2}{2^{2n}}$$

□

Theorem 6.5 *Let $f, h \in_r F_n$ be independently chosen pseudorandom functions, and $G_2 = \psi(h, 1, f, h, 1)$ and $G_3 = \psi(f, 1, h, f, 1)$. When*

$$| \text{Prob}\{DC_{2n}[G_2, G_3] = 1\} - \text{Prob}\{DC_{2n}[P_{2n}, P_{2n}] = 1\} | < \frac{1}{n^{c_2}}$$

for any polynomial size D-distinguishing circuit and for any constant c_2, then $G_1 = \psi(h, 1, f, h, 1, f)$ is a super-pseudorandom permutation generator.

Proof :   First it is necessary to show that $G_3 = \psi(f, 1, h, f, 1)$ and $\overline{G}_1 = \psi(f, 1, h, f, 1, h)$ are independent of each other. In order to prove this, assume that they are not independent, and that there is a D-distinguishing circuit such that for a constant c_2

$$| \text{Prob}\{DC_{2n}[G_3, \overline{G}_1] = 1\} - \text{Prob}\{DC_{2n}[P_{2n}, P_{2n}] = 1\} | > \frac{1}{n^{c_2}}$$

Without changing the inequality relation, we have

$$| \text{Prob}\{DC_{2n}[G_3, \overline{G}_1] = 1\} - \text{Prob}\{DC_{2n}[G_3, G_3] = 1\} +$$
$$\text{Prob}\{DC_{2n}[G_3, G_3] = 1\} - \text{Prob}\{DC_{2n}[P_{2n}, P_{2n}] = 1\} | > \frac{1}{n^{c_2}}$$

Then

$$| \text{Prob}\{DC_{2n}[G_3, \overline{G}_1] = 1\} - \text{Prob}\{DC_{2n}[G_3, G_3] = 1\} | +$$
$$| \text{Prob}\{DC_{2n}[G_3, G_3] = 1\} - \text{Prob}\{DC_{2n}[P_{2n}, P_{2n}] = 1\} | > \frac{1}{n^{c_2}}$$

If $| \text{Prob}\{DC_{2n}[G_3, G_3] = 1\} - \text{Prob}\{DC_{2n}[P_{2n}, P_{2n}] = 1\} | > \frac{1}{n^{c_2}}$, G_3 is not pseudorandom, as the D-distinguishing circuit is not a test for identity. This contradicts Lemma 6.2. Furthermore, if $| \text{Prob}\{DC_{2n}[G_3, \overline{G}_1] =$

$1\} - \text{Prob}\{DC_{2n}[G_3, G_3] = 1\} \mid > \frac{1}{n^{c_2}}$, then the oracle circuit virtually distinguishes f from a randomly chosen function. This also contradicts our assumption that f is a pseudorandom function. Since both cases reduce to contradictions, then G_3 and $\overline{G}_1$ must be independent of each other, and there is no D-distinguishing circuit for $(\overline{G}_1, G_3)$.

Considering the independence of G_2 and G_3, we have that

$$\mid \text{Prob}\{DC_{2n}[G_2, G_3] = 1\} - \text{Prob}\{DC_{2n}[P_{2n}, P_{2n}] = 1\} \mid < \frac{1}{n^{c_2}}$$

for any constant c_2. Note that G_2 and G_3 are not inverse to each other; so the D-distinguishing circuit cannot be a test for inversion. Without changing the sign of inequality, the above relation can be expanded as,

$$\mid \text{Prob}\{DC_{2n}[G_1, \overline{G}_1] = 1\} - \text{Prob}\{DC_{2n}[G_1, \overline{G}_1] = 1\} \; +$$
$$\text{Prob}\{DC_{2n}[G_3, \overline{G}_1] = 1\} - \text{Prob}\{DC_{2n}[G_3, \overline{G}_1] = 1\} \; +$$
$$\text{Prob}\{DC_{2n}[G_2, G_3] = 1\} - \text{Prob}\{DC_{2n}[P_{2n}, P_{2n}] = 1\} \mid \; < \; \frac{1}{n^{c_2}}$$

Then

$$\mid\mid \text{Prob}\{DC_{2n}[G_2, G_3] = 1\} - \text{Prob}\{DC_{2n}[G_3, \overline{G}_1] = 1\} \mid \; -$$
$$\mid \text{Prob}\{DC_{2n}[G_3, \overline{G}_1] = 1\} - \text{Prob}\{DC_{2n}[G_1, \overline{G}_1] = 1\} \mid \; -$$
$$\mid \text{Prob}\{DC_{2n}[G_1, \overline{G}_1] = 1\} - \text{Prob}\{DC_{2n}[P_{2n}, P_{2n}] = 1\} \mid\mid \; < \; \frac{1}{n^{c_2}}$$

Since it was assumed that G_2 and G_3 are independent permutations, and so are G_3 and $\overline{G}_1$, then $\mid \text{Prob}\{DC_{2n}[G_2, G_3] = 1\} - \text{Prob}\{DC_{2n}[G_3, \overline{G}_1] = 1\} \mid$ would be less than $\frac{1}{n^c}$ for any constant c, since

$$\mid \text{Prob}\{DC_{2n}[G_2, G_3] = 1\} - \text{Prob}\{DC_{2n}[G_3, \overline{G}_1] = 1\} \mid \; \leq$$
$$\mid \text{Prob}\{DC_{2n}[G_2, G_3] = 1\} - \text{Prob}\{DC_{2n}[P_{2n}, P_{2n}] = 1\} \mid \; +$$
$$\mid \text{Prob}\{DC_{2n}[G_3, \overline{G}_1] = 1\} - \text{Prob}\{DC_{2n}[P_{2n}, P_{2n}] = 1\} \mid \; < \; \frac{1}{n^c}$$

Hence

$$\mid \text{Prob}\{DC_{2n}[G_1, \overline{G}_1] = 1\} - \text{Prob}\{DC_{2n}[P_{2n}, P_{2n}] = 1\} \mid \; +$$
$$\mid \text{Prob}\{DC_{2n}[G_3, \overline{G}_1] = 1\} - \text{Prob}\{DC_{2n}[G_1, \overline{G}_1] = 1\} \mid \; < \; \frac{1}{n^{c_2}}$$

So, each of the above absolute values is less than $\frac{1}{n^{c_2}}$. In other words

$$\mid \text{Prob}\{DC_{2n}[G_1, \overline{G}_1] = 1\} - \text{Prob}\{DC_{2n}[P_{2n}, P_{2n}] = 1\} \mid < \frac{1}{n^{c_2}}$$

Hence G_1 and $\overline{G}_1$ are independent of each other, and G_1 is a super-pseudorandom permutation.   □

Lemmas 6.1, 6.2 and 6.3 and Theorem 6.5 show that, when the number of oracle gates m is a polynomial in n, then the probability of making a circuit which distinguishes $G_2 = \psi(h, 1, f, h, 1)$ from $G_3 = \psi(f, 1, h, f, 1)$ is less than $\frac{1}{n^{c_2}}$, for any constant c_2, and sufficiently large n. Furthermore, the probability of making a super-distinguishing circuit for $G_1 = \psi(h, 1, f) \circ \psi(h, 1, f)$ is also less than $\frac{1}{n^{c_2}}$ for any constant c_2, and sufficiently large n. In other words, G_1 is secure against chosen plaintext/ciphertext attack if the cryptanalyst is permitted to make only a polynomial number of queries.

6.4.2   Super-Pseudorandomness of $\psi(f^2, 1, f, f^2, 1, f)$

In Lemmas 6.1, 6.2 and 6.3, upper bounds on the probabilities of distinguishing between two permutation generators have been found, where f and h are two independently chosen pseudorandom functions. In the following lemmas and theorem, we show that if f^2 is substituted for h, there is an increase in the upper bound on the above probabilities. Nevertheless, if the number of oracles is limited to some polynomial in n, the corresponding probabilities would remain less than $\frac{1}{n^{c_2}}$ for any constant c_2, and sufficiently large n.

Lemma 6.4 *Let $f, h \in_r H_n$ and let C_{2n} be a distinguishing circuit with $m < 2^n$ oracle gates; then*

$$| \operatorname{Prob}\{C_{2n}[\psi(f^2, 1, f, f^2, 1)] = 1\} - \operatorname{Prob}\{C_{2n}[\psi(h, 1, f, h, 1)] = 1\} | \leq \frac{2m^2}{2^n}$$

Proof :   Since both f and h can be considered to be two sequences of 2^n independent and uniformly distributed n-bit random variables, for an argument $a \in \Sigma^n$, $f(a)$ and $h(a)$ are two independent n-bit strings. When the input to an oracle is $(L \parallel R)$ and the oracle is evaluated with $\psi(h, 1, f, h, 1)$, each branch of the outputs (that is, S and T) is always a sum of two random variables generated by the functions f and h (see Figure 6.9). Thus

$$S = L \oplus R \oplus f(L \oplus h(L \oplus R)) \oplus h(R \oplus h(L \oplus R) \oplus f(L \oplus h(L \oplus R)))$$
$$T = R \oplus h(L \oplus R) \oplus f(L \oplus h(L \oplus R))$$

where all the random variables $Y_i = f(L_i \oplus h(L_i \oplus R_i))$ are independent of

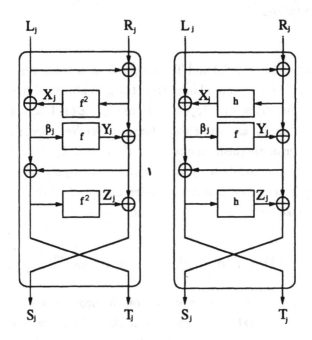

Figure 6.9: Random Variables X, Y and Z in $\mathcal{G}_2$ and G_2

the random variables

$$X_j = h(L_j \oplus R_j)$$
$$Z_j = h(R_j \oplus h(L_j \oplus R_j) \oplus f(L_j \oplus h(L_j \oplus R_j)))$$

When the input to an oracle is $(L \parallel R)$ and the oracle is evaluated with $\psi(f^2, 1, f, f^2, 1)$, each branch of the output (that is, S and T) is always a sum of two random variables generated by the function f. Thus

$$S = L \oplus R \oplus f(L \oplus f^2(L \oplus R)) \oplus f^2(R \oplus f^2(L \oplus R) \oplus f(L \oplus f^2(L \oplus R)))$$
$$T = R \oplus f^2(L \oplus R) \oplus f(L \oplus f^2(L \oplus R))$$

If $R = 0$, then

$$S = L \oplus f(L \oplus f^2(L)) \oplus f^2(f^2(L) \oplus f(L \oplus f^2(L)))$$
$$T = f^2(L) \oplus f(L \oplus f^2(L))$$

If $L = 0$, then

$$S = R \oplus f^3(R)) \oplus f^2(R \oplus f^2(R) \oplus f^3(R))$$
$$T = R \oplus f^2(R) \oplus f^3(R)$$

If $L = 0$ and $R = 0$, then

$$
\begin{aligned}
S &= f^3(0) \oplus f^2(f^2(0) \oplus f^3(0)) \\
T &= f^2(0) \oplus f^3(0)
\end{aligned}
$$

When all the oracle gates are evaluated with $\psi(f^2, 1, f, f^2, 1)$ or all the oracle gates are evaluated with $\psi(h, 1, f, h, 1)$, a distinguisher generates a bit **1** on its output with the same probability if all the random variables $Y_i = f(L_i \oplus f^2(L_i \oplus R_i))$ are independent of the random variables

$$
\begin{aligned}
X_j &= f^2(L_j \oplus R_j) \\
Z_j &= f^2(R_j \oplus f^2(L_j \oplus R_j) \oplus f(L_j \oplus f^2(L_j \oplus R_j)))
\end{aligned}
$$

In other words, a distinguisher with m oracle gates generates a bit **1** as its output when there is one oracle gate O_i such that β_i, the input value to the random function f, is equal to either of

$$
\beta_i = \begin{cases} f(L_j \oplus R_j) \\ f(R_j \oplus f^2(L_j \oplus R_j) \oplus f(L_j \oplus f^2(L_j \oplus R_j))) \end{cases}
$$

for some $j = 1, \ldots, m$. The probability that in a given oracle the input to the f function takes a value equal to any of the $2m$ internal random values in m oracle gates with different inputs is $\frac{2m}{2^n}$. The probability that a circuit distinguishes $\psi(f^2, 1, f, f^2, 1)$ from $\psi(h, 1, f, h, 1)$ is equal to the probability that two of the oracles generate dependent random variables. Hence

$$
\mid \mathrm{Prob}\{C_{2n}[\psi(f^2, 1, f, f^2, 1)] = 1\} - \mathrm{Prob}\{C_{2n}[\psi(h, 1, f, h, 1)] = 1\} \mid \leq \frac{2m^2}{2^n}
$$

The probability of distinguishing $\psi(h, 1, f, h, 1)$ from a random permutation was given in Lemma 6.1. As a result, an upper bound on the overall probability of distinguishing $\mathcal{G}_2 = \psi(f^2, 1, f, f^2, 1)$ from a random permutation is

$$
\mid \mathrm{Prob}\{C_{2n}[\mathcal{G}_2] = 1\} - \mathrm{Prob}\{C_{2n}[P_{2n}] = 1\} \mid \leq \frac{3m^2}{2^n} + \frac{m^2}{2^{2n}}
$$

where $m < 2^n$ is the number of oracle gates. Hence, when the oracle circuit is bounded by a polynomial number of oracle gates, the probability of distinguishing $\psi(f^2, 1, f, f^2, 1)$ from a random permutation is less than $\frac{1}{n^{c_2}}$ for any constant c_2, and for sufficiently large n. $\square$

Lemma 6.5 *Let $f, h \in_r H_n$ and let C_{2n} be a distinguishing circuit with $m < 2^n$ oracle gates; then*

$$| \operatorname{Prob}\{C_{2n}[\psi(f, 1, f^2, f, 1)] = 1\} - \operatorname{Prob}\{C_{2n}[\psi(f, 1, h, f, 1)] = 1\} | \leq \frac{2m^2}{2^n}$$

Proof :   The proof is similar to the proof of Lemma 6.4. When the input to an oracle is $(L \parallel R)$ and the oracle is evaluated with $\psi(f, 1, f^2, f, 1)$, each branch of the output (that is S and T) is always a sum of two random variables generated by the function f. Thus

$$\begin{aligned}
S &= L \oplus R \oplus f^2(L \oplus f(L \oplus R)) \oplus f(R \oplus f(L \oplus R) \oplus f^2(L \oplus f(L \oplus R))) \\
T &= R \oplus f(L \oplus R) \oplus f^2(L \oplus f(L \oplus R))
\end{aligned}$$

If $R = 0$, then

$$\begin{aligned}
S &= L \oplus f^2(L \oplus f(L)) \oplus f(f(L) \oplus f^2(L \oplus f(L))) \\
T &= f(L) \oplus f^2(L \oplus f(L))
\end{aligned}$$

If $L = 0$, then

$$\begin{aligned}
S &= R \oplus f^3(R)) \oplus f(R \oplus f(R) \oplus f^3(R)) \\
T &= R \oplus f(R) \oplus f^3(R)
\end{aligned}$$

If $L = 0$ and $R = 0$, then

$$\begin{aligned}
S &= f^3(0) \oplus f(f(0) \oplus f^3(0)) \\
T &= f(0) \oplus f^3(0)
\end{aligned}$$

When all the oracle gates are evaluated with $\psi(f, 1, f^2, f, 1)$ or all the oracle gates are evaluated with $\psi(f, 1, h, f, 1)$, a distinguisher generates a bit **1** as its output with the same probability, if all the random variables $Y_i = f^2(L_i \oplus f(L_i \oplus R_i))$ are independent of the random variables

$$\begin{aligned}
X_j &= f(L_j \oplus R_j) \\
Z_j &= f(R_j \oplus f(L_j \oplus R_j) \oplus f^2(L_j \oplus f(L_j \oplus R_j)))
\end{aligned}$$

In other words, a distinguisher with m oracle gates generates **1** as its output when there is one oracle gate O_i such that β_i, the input value to the random function f, is equal to $f(L_j \oplus f(L_j \oplus R_j))$ for some $j = 1, \ldots, m$. Since f

was used in two different layers for each oracle, the probability that, in a given oracle β_i, the input to the f functions takes a value equal to any of m internal random values Y_i, that is, the output of the f^2 layer in m oracle gates with different inputs, is $\frac{2m}{2^n}$. The probability that a circuit distinguishes $\psi(f, 1, f^2, f, 1)$ from $\psi(f, 1, h, f, 1)$ is equal to the probability that two of the oracles generate dependent random variables. Hence

$$| \operatorname{Prob}\{C_{2n}[\psi(f,1,f^2,f,1)] = 1\} - \operatorname{Prob}\{C_{2n}[\psi(f,1,h,f,1)] = 1\} | \leq \frac{2m^2}{2^n}$$

The probability of distinguishing $\psi(f, 1, h, f, 1)$ from a random permutation was given in Lemma 6.2. As a result, an upper bound on the overall probability of distinguishing $\mathcal{G}_3 = \psi(f, 1, f^2, f, 1)$ from a random permutation is

$$| \operatorname{Prob}\{C_{2n}[\mathcal{G}_3] = 1 - \operatorname{Prob}\{C_{2n}[P_{2n}] = 1\} | \leq \frac{3m^2}{2^n} + \frac{m^2}{2^{2n}}$$

where $m \leq 2^n$ is the number of oracle gates. Hence, when the oracle circuit is bounded by a polynomial number of oracle gates, the probability of distinguishing $\psi(f, 1, f^2, f, 1)$ from a random permutation is less than $\frac{1}{n^{c_2}}$ for any constant c_2, and sufficiently large n. $\square$

Theorem 6.6 *Let $f \in_r F_n$ be a pseudorandom function; then*

$$\mathcal{G}_1 = \psi(f^2, 1, f, f^2, 1, f) ,$$

is a super-pseudorandom permutation.

Proof : To prove that $\mathcal{G}_1$ is a super-pseudorandom permutation generator, we first show that $\mathcal{G}_2 = \psi(f^2, 1, f, f^2, 1)$ and $\mathcal{G}_3 = (f, 1, f^2, f, 1)$ are independent of each other. As was shown in Lemma 6.4 and Lemma 6.5, in a D-distinguishing circuit when the input to an oracle is $(L \parallel R)$ and the oracle is evaluated with $\psi(f^2, 1, f, f^2, 1)$, the output is

$$S = L \oplus R \oplus f(L \oplus f^2(L\oplus R)) \oplus f^2(R \oplus f^2(L\oplus R) \oplus f(L \oplus f^2(L\oplus R)))$$
$$T = R \oplus f^2(L \oplus R) \oplus f(L \oplus f^2(L \oplus R))$$

When the oracle is evaluated with $\psi(f, 1, f^2, f, 1)$, the output is

$$S' = L \oplus R \oplus f^2(L \oplus f(L \oplus R)) \oplus f(R \oplus f(L \oplus R) \oplus f^2(L \oplus f(L \oplus R)))$$
$$T' = R \oplus f(L \oplus R) \oplus f^2(L \oplus f(L \oplus R))$$

The six random variables involved are:

$$X_2 = f^2(L \oplus R)$$
$$Y_2 = f(L \oplus f^2(L \oplus R))$$
$$Z_2 = f^2(R \oplus f^2(L \oplus R) \oplus f(L \oplus f^2(L \oplus R)))$$

and

$$X_3 = f(L \oplus R)$$
$$Y_3 = f^2(L \oplus f(L \oplus R))$$
$$Z_3 = f(R \oplus f(L \oplus R) \oplus f^2(L \oplus f(L \oplus R)))$$

S, T, S' and T' are always a sum of two random variables generated by the function f. If a random variable in a output branch of an oracle becomes dependent on a random variable in any output branch of another oracle, the other random variables are always independent of each other. For example, if X_2 is equal to X_3, the probability that Y_2 is equal to Y_3 is $\frac{1}{2^n}$. Likewise, if $Y_2 = Y_3$, the probability that $Z_2 = Z_3$ is $\frac{1}{2^n}$. Hence, the probability of dependence between two branches is equal to $\frac{1}{2^n}$, which is equal to the probability of dependence between two output branches in two different oracle gates when, instead of f^2, an independent random function such as h is applied. Here, we calculate an upper bound on the probability of independence. As was shown earlier in Lemma 6.4, $\psi(f^2, 1, f, f^2, 1)$ and $\psi(h, 1, f, h, 1)$ are indistinguishable from each other; it was shown in Theorem 6.5 that $\psi(h, 1, f, h, 1)$ and $\psi(f, 1, h, f, 1)$ are independent of each other, and it was also shown in Lemma 6.5 that $\psi(f, 1, h, f, 1)$ and $\psi(f, 1, f^2, f, 1)$ are indistinguishable from each other. The probability that there is a D-distinguishing circuit for $\mathcal{G}_2$ and $\mathcal{G}_3$ can be written as:

$$\mid \text{Prob}\{DC_{2n}[\mathcal{G}_2, \mathcal{G}_3] = 1\} - \text{Prob}\{DC'_{2n}[P_{2n}, P_{2n}] = 1\} \mid < \frac{1}{n^{c_2}}$$

This statement can be rewritten as,

$$\mid \text{Prob}\{DC_{2n}[\mathcal{G}_2, \mathcal{G}_3] = 1\} \quad - \quad \text{Prob}\{DC_{2n}[\mathcal{G}_2, G_3] = 1\}$$
$$+ \text{Prob}\{DC_{2n}[\mathcal{G}_2, G_3] = 1\} \quad - \quad \text{Prob}\{DC_{2n}[G_2, G_3] = 1\}$$
$$+ \text{Prob}\{DC_{2n}[G_2, G_3] = 1\} \quad - \quad \text{Prob}\{DC_{2n}[P_{2n}, P_{2n}] = 1\} \mid < \frac{1}{n^{c_2}}$$

where $G_2 = \psi(h, 1, f, h, 1)$ and $G_3 = \psi(f, 1, h, f, 1)$. With reordering and separation of absolute values, we get

$$
\begin{aligned}
\| \operatorname{Prob}\{DC_{2n}[\mathcal{G}_2, \mathcal{G}_3] = 1\} &- \operatorname{Prob}\{DC_{2n}[\mathcal{G}_2, G_3] = 1\} \, | \\
+ \, | \operatorname{Prob}\{DC_{2n}[\mathcal{G}_2, G_3] = 1\} &- \operatorname{Prob}\{DC_{2n}[G_2, G_3] = 1\} \, | \\
+ \, | \operatorname{Prob}\{DC_{2n}[G_2, G_3] = 1\} &- \operatorname{Prob}\{DC_{2n}[P_{2n}, P_{2n}] = 1\} \, \|
\end{aligned}
$$

If the sum of the above probabilities is less than 1 over any polynomial in n, then $\mathcal{G}_2$ and $\mathcal{G}_3$ are essentially independent of each other according to the definition. When f and $h \in_r H_n$, by applying procedures similar to those of the proofs of Lemma 6.4, Lemma 6.5 and Lemma 6.2, it can be shown that the first term is less than $\frac{2m^2}{2^n}$, the second term is less than $\frac{2m^2}{2^n}$, and the third term is less than $\frac{m^2}{2^n} + \frac{m^2}{2^{2n}}$, respectively. So, the sum of these three probabilities, gives a bound on the probability for making a D-distinguishing circuit with m oracle gates for $\mathcal{G}_2$ and $\mathcal{G}_3$ as $\frac{5m^2}{2^n} + \frac{m^2}{2^{2n}}$. When there is a polynomial number of oracle gates, that is, m is a polynomial in n, then

$$
| \operatorname{Prob}\{DC_{2n}[\mathcal{G}_3, \mathcal{G}_2] = 1\} - \operatorname{Prob}\{DC_{2n}[P_{2n}, P_{2n}] = 1\} \, | < \frac{1}{n^{c_2}}
$$

for any constant c_2, and sufficiently large n. By applying a proof similar to the proof of Theorem 6.5, it can be shown that, when $\mathcal{G}_2$ and $\mathcal{G}_3$ are independent, $\mathcal{G}_1$ is super-pseudorandom.   □

6.5   Conclusions and Open Problems

We have shown that it is possible to construct a super-pseudorandom permutation generator by applying a single pseudorandom function. We took advantage of the structure for an optimal perfect randomizer presented in [Pieprzyk and Sadeghiyan, 1991]. We first showed that $\psi(h, 1, f, h, 1, f)$ is a super-pseudorandom permutation. Then we showed that, by substituting f^2 for h, the probability of making a super-distinguisher is still less than 1 over any polynomial in n. As $\psi(f^2, 1, f, f^2, 1, f)$ is super-pseudorandom, it can be applied as a block cipher secure against chosen plaintext/ciphertext attack. Although such a block cryptosystem is less than practical, it can be viewed as an attempt towards the construction of practical ones which are provably secure against chosen plaintext/ciphertext attack without relying on any unproven hypothesis. However, in the same way that the results of Luby and Rackoff were considered to be a justification for the application of DES-like permutations in the design of DES, the above structure can be adopted in the design of block ciphers to yield stronger cryptographic properties. If a cryptosystem is super-pseudorandom, it can be applied in block-cipher-based hash schemes. Two open problems emerge from the results of this chapter. The structure applies 6 rounds of DES-like permutations. The first open problem is whether the proposed structure is optimal, and whether any other structure can be suggested which needs fewer rounds. We do not know whether $\psi(f, f^2, f, f, f)$ is such a structure. The second open problem is whether the proposed structure can be adopted to improve the quality of existing cryptosystems such as DES or LOKI against differential cryptanalysis without needing to redesign their S-boxes.

Chapter 7

A Construction for One Way Hash Functions and Pseudorandom Bit Generators

7.1 Introduction

In Chapter 2, we listed the properties that a secure hash function should satisfy, among them was the property of one-wayness. Several approaches to constructing hash functions have applied DES, or other block ciphers such as LOKI, as the underlying one-way function. Unfortunately, DES suffers from a small key space and also has other undesired properties such as the complementation property. In Chapters 4, 5 and 6 we developed a structure to be employed in the design of block ciphers used in block-cipher-based hash schemes. On the other hand, block ciphers are not the only functions which are considered to be one-way and difficult to invert. For example, functions such as RSA or the squaring modulo composite N are considered to be one-way. In Chapters 7 and 8 we develop some generalized constructions for hash functions from one-way permutations.

The current trend in cryptography is to provide the construction of basic primitives with general cryptographic assumptions that are as weak as possible. It is theoretically important to base cryptographic primitives and basic tools on reduced complexity assumptions. Practically it is important to give efficient implementations of such constructions. Each successive paper on the construction of hash functions has assumed weaker conditions for the one-way function, and then suggested a construction for hashing with more complicated procedures. Finally, Rompel gave a construction for one-way hash functions from any one-way function and proved that the existence of one-way functions is a necessary and sufficient condition for constructing hash functions [Rompel, 1990]. However, although his work is theoretically optimal, it is less than practical. We give a brief description of these theoretic constructions later in Section 7.4.

Two formal complexity-theoretic definitions have been suggested for cryptographic hash function families. The first family of hash functions, defined by Damgard, is the Collision Free Hash Functions (CFHF) or Collision Intractable Hash Functions (CIH). We gave a rough definition of this type of hash functions in Section 2.3.1, where they were called strong hash functions. We give the precise formal definition of such a family of functions in Section 7.3. The second family, defined by Naor and Yung, is the Universal One Way Hash Functions (UOWHF). This family is weaker than the previous one. We gave a rough definition for this type of hash functions in Section 2.3.1, where they were called weak hash functions. We will also give the precise formal definition of such family of functions in Section 7.3.

Zheng, Matsumoto and Imai revealed a duality between pseudorandom bit generators (PBG) and UOWHF. Applying the revealed duality, they presented a construction for UOWHF which is equivalent to the construction of Blum-Micali pseudorandom bit generators [Zheng et al., 1990a]. Blum and Micali discovered hard-core predicates b of functions f [Blum and Micali, 1984]. Such predicates cannot be efficiently obtained, given $f(x)$. They applied this notion to construct a PBG based on the intractability of the discrete logarithm problem. However, the efficiency of this method is limited by the number of *hard bits* of the underlying one-way permutation. It is noteworthy that Yao generalized this scheme by showing that a PBG can be constructed from any one-way permutation [Yao, 1982]. He transforms any one-way permutation into a more complicated one which has a hard-core predicate. Similar to the works on hash functions, later works on the construction of pseudorandom bit generators have tried to make more generalizations and assume weaker conditions on the one-way function, for example see [Goldreich and Levin, 1989] and [Impagliazzo *et al.*, 1989].

In this chapter, we present a method such that given an n-bit one-way permutation with some known hard bits, a one-way permutation with n hard bits can be constructed. We call this one-way permutation a strong permutation. We apply it to present a construction for pseudorandom bit generators with maximum efficiency, based on the Blum-Micali pseudorandom bit generator. We also present a method to build a universal one-way hash function from the strong permutation. Hence, given a one-way permutation, we can construct both an efficient pseudorandom generator and a universal one-way hash function. We show that by the application of the strong permutation, Zheng, Matsumoto and Imai's scheme can be reduced to Damgard's design principle for construction of hash functions, and will yield the same result. Therefore, our proposal yields an algorithm that can be used both for generating pseudorandom bits, and hashing long messages. This has a practical significance, since it would not be necessary to use two different algorithms for implementing these two cryptographic tools. The results of this chapter have appeared in [Sadeghiyan and Pieprzyk, 1991a].

7.2   Notation

The notation we use in this chapter and Chapter 8 is similar to [Zheng *et al.*, 1990a]. The set of all non-negative integers is denoted by N. Let $\Sigma = \{0, 1\}$ to be the alphabet we consider. For $n \in N$, Σ^n is the set of all binary strings of length n. The concatenation of two binary strings x, y is denoted by $x \parallel y$. The length of a string x is denoted by $\mid x \mid$.

Let l be a monotone increasing function from N to N and f a function[1] from D to R, where $D = \bigcup_n D_n$, $D_n \subseteq \Sigma^n$ and $R = \bigcup_n R_n$, $R_n \subseteq \Sigma^{l(n)}$. D is called the domain and R the range of f. Denote by f_n the restriction of f to Σ^n. The function f is a permutation if each f_n is a one-to-one and onto function. f is polynomial time computable if there is a polynomial time algorithm computing $f(x)$ for all $x \in D$. The composition of two functions f and g is defined as $f \circ g(x) = f(g(x))$. The i-fold composition of f is denoted by $f^{(i)}$.

---

[1]Note that the definition of function in this chapter is different from the definition of function in Section 4.5.

An (probability) ensemble E, with length $l(n)$, is a family of probability distributions $\{E_n \mid E_n : \Sigma^{l(n)} \to [0,1], n \in N\}$. The uniform ensemble U with length $l(n)$ is the family of uniform probability distributions U_n, where each U_n is defined as $U_n(x) = \frac{1}{2^{l(n)}}$, for all $x \in \Sigma^{l(n)}$. By $x \in_E \Sigma^{l(n)}$ we mean that x is randomly selected from $\Sigma^{l(n)}$ according to E_n, and by $x \in_r S$ we mean that x is chosen from the set S uniformly at random. E is samplable if there is an algorithm M that given input n, outputs an $x \in_E \Sigma^{l(n)}$, and polynomially samplable if the running time of M is also polynomially bounded.

7.3 Preliminaries

In this section, the formal definitions for Universal One Way Hash Functions and Collision Free Hash Functions together with some preliminary definitions are presented, which are used throughout Chapters 7 and 8. We give other required definitions as necessary in the text.

Definition 7.1 *A statistical test is a probabilistic algorithm T that given an input x, where x is an n-bit string, halts in time $O(n^t)$ and outputs a bit 0 or 1, where t is some fixed positive integer.*

Definition 7.2 *Let l be a polynomial, and E^1 and E^2 be ensembles both with length $l(n)$. E^1 and E^2 are called* indistinguishable *from each other, if for each statistical test T, for each polynomial Q and for all sufficiently large n,*

$$|\ Prob\{T(x_1) = 1\} - Prob\{T(x_2) = 1\}\ | < \frac{1}{Q(n)}$$

where $x_1 \in_{E^1} \Sigma^{l(n)}$, $x_2 \in_{E^2} \Sigma^{l(n)}$.

Definition 7.3 *A polynomially samplable ensemble E is* pseudorandom *if it is indistinguishable from the uniform ensemble U with the same length.*

Definition 7.4 *Let $f : D \to R$, where $D = \bigcup_n \Sigma^n$ and $R = \bigcup_n \Sigma^{l(n)}$, be a polynomial time computable function. We say that f is one-way if for each* *probabilistic polynomial time algorithm M, for each polynomial Q and for all sufficiently large n,*

$$\text{Prob}\{f_n(M(f_n(x))) = f_n(x)\} < \frac{1}{Q(n)}$$

where $x \in_U D_n$.

Let l be a polynomial with $l(n) > n$ and let H be a family of hash functions defined by $H = \bigcup_n H_n$, where H_n is a set of functions from $\Sigma^{l(n)}$ to Σ^n. For two strings $x, y \in \Sigma^{l(n)}$ with $x \neq y$, we say that x and y collide under $h \in H_n$ or (x, y) is a collision pair for h, if $h(x) = h(y)$. H is polynomial time computable if there is a polynomial time algorithm computing all $h \in H$, and accessible if there is a probabilistic polynomial time algorithm that on input $n \in N$ outputs uniformly at random a description of $h \in H_n$. Let F be a *collision finder*. F is a probabilistic polynomial time algorithm such that on input $x \in \Sigma^{l(n)}$ and $h \in H_n$ outputs either ? (cannot find) or a string $y \in \Sigma^{l(n)}$ such that $x \neq y$ and $h(x) = h(y)$. The definition for a Universal One Way Hash Functions (UOWHF) is formally described as follows.

Definition 7.5 *Let H be a computable and accessible hash function compressing $l(n)$-bit input into n-bit output strings and F a collision string finder. H is a universal one-way hash function if for each F, and for each polynomial Q and for all sufficiently large n,*

$$\text{Prob}\{F(x, h) \neq ?\} < \frac{1}{Q(n)}$$

where $x \in \Sigma^{l(n)}$ and $h \in_r H_n$. The probability is computed over all $h \in_r H_n$, $x \in \Sigma^{l(n)}$ and the random choice of all finite strings that F could have chosen.

The definition for Collision Free Hash Function is given by Damgard in [Damgard, 1987]. Let A be a *collision-pair finder*. A is a probabilistic polynomial time algorithm that on input $h \in H_n$ outputs either ? or a pair of strings $x, y \in \Sigma^{l(n)}$ with $x \neq y$ and $h(x) = h(y)$. The definition for a Collision Free Hash Function is formally described as follows:

Definition 7.6 *H is called a collision free hash function if for each A, and for each polynomial Q, and for all sufficiently large n,*

$$\text{Prob}\{A(h) \neq ?\} < \frac{1}{Q(n)}$$

where $h \in_r H_n$, *and the probability* $\text{Prob}\{A(h) \neq ?\}$ *is computed over all* $h \in H_n$ *and the random choice of all finite strings that A could have chosen.*

Note that a particular hash scheme was considered to be secure in Chapter 2, if there was no algorithm which could find colliding messages using the computing resources that today's technology provides. But a theoretical construction is considered to be secure if there is no algorithm running in polynomial time which can find colliding messages.

7.4   Theoretic Constructions

In this section we give a brief review of constructions suggested for Universal One Way Hash Functions, which have more theoretic significance. Naor and Yung were the first who introduced the concept of a UOWHF and suggested a construction based on a one-way permutation. In their construction, they took advantage of the notion of *universal family of functions with collision accessibility property*, which had already been introduced in [Carter and Wegman, 1979]. The general definition for these functions is given in the following:

Definition 7.7 *Suppose G is a set of functions and each element of G is a function from A to B. G is* strongly universal$_r$ *if given any r distinct elements* $a_1, \ldots, a_r$ *of A, and any r elements* $b_1, \ldots, b_r$ *of B, then there are* $\frac{(\#G)}{(\#B)^2}$ *functions which take* a_1 *to* b_1 *and* a_2 *to* b_2 *and so on.*

Definition 7.8 *A strongly universal$_r$ family of functions G has the* collision accessibility property *if it is possible to generate in polynomial time a function* $g \in G$ *that obeys the requirement*

$$g(a_1) = b_1$$
$$g(a_2) = b_2$$
$$\vdots$$
$$g(a_r) = b_r$$

7.4.1   Naor and Yung's Scheme

Naor and Yung showed that the existence of a secure signature scheme reduces to the existence of a UOWHF [Naor and Yung, 1989]. Furthermore, the construction of a UOWHF can be reduced to the construction of a UOWHF that compresses one bit. Using the same method as introduced by Carter and Wegman in [Carter and Wegman, 1979] and [Wegman and Carter, 1981], Naor and Yung constructed a family of UOWHF's by the composition of any one-way permutation and a family of strongly universal$_2$ hash functions with the collision accessibility property. The general definition for strongly universal$_r$ functions is given in Definition 7.7. In Naor and Yung's construction, the one-way permutation provides the one-wayness of the UOWHF, and the strongly universal$_2$ family of hash functions performs the mapping to the small length output. When a member is chosen randomly and uniformly from the family, the output is distributed randomly and uniformly over the output space.

Theorem 7.1 *Let f be a one-way permutation on Σ^n and let G_n be a strongly* universal$_2$ *family of hash functions from Σ^n to Σ^{n-1}, then $H_n = \{h = g \circ f \mid g \in G_n\}$ is a UOWHF compressing n-bit input strings to $(n-1)$-bit output strings.*

In the above construction, the size of the description of the hash function is $O(n^2)$, where n is the size of the input. The above construction is not efficient in practice, as it compresses only one bit each time. This can be improved by a factor t when a strongly universal$_t$ hash function is used.

7.4.2   Zheng, Matsumoto and Imai's First Scheme

This construction is based on the composition of a pairwise independent uniformizer and a strongly universal hash function with a quasi-injection one-way function[2]. This construction together with the definition of a pairwise independent uniformizer is given below:

---

[2]As the definition of quasi-injection one-way functions involves other definitions which are beyond the scope of this brief survey, we refer an interested reader to [Zheng *et al.*, 1990b].

Definition 7.9 *Let V_n be a set of permutations from $l(n)$-bit strings to $l(n)$-bit strings. $V = \bigcup_n V_n$ is a pairwise independent uniformizer, if for each n, for every (x_1, x_2) and for every (y_1, y_2), with $x_1, x_2, y_1, y_2 \in \Sigma^n$ and $x_1 \neq x_2$ and $y_1 \neq y_2$, there are exactly*

$$\frac{(\#V_n)}{2^{l(n)}(2^{l(n)} - 1)}$$

permutations in V_n that map x_1 to y_1 and x_2 to y_2.

Theorem 7.2 *Let f be a quasi injection one-way function from D to R, where $D = \bigcup_n \Sigma^n$, $R = \bigcup_n \Sigma^{m(n)}$ and let m be a polynomial with $m(n) \geq n$. Let $V = \bigcup_n V_n$ be a pairwise independent uniformizer with length $m(n)$ and let $G = \bigcup_n G_n$ be a strongly universal hash function that compresses $m(n)$-bit input into $(n-1)$-bit output strings with the collision accessibility property. Then $H_n = \{h \mid h = g \circ v \circ f_{n+1}, g \in G_{n+1}, v \in V_{n+1}\}$, is a universal one-way hash function compressing $(n+1)$-bit input into n-bit output strings.*

7.4.3 De Santis and Yung's Schemes

De Santis and Yung made two contributions in this area. First, they improved the construction of Naor and Yung by applying a one-to-one one-way function, instead of a one-way permutation. Second, they presented two constructions for a UOWHF with weaker assumptions on the applied one-way function [De Santis and Yung, 1990].

The first construction is based on the existence of a one-way function with small expected preimage size, which is a one-way function such that, when an element in the domain is chosen randomly, the expected size of the preimage of the element in the range is small. An example of such a function is squaring modulo a composite. Another example is any one-way function which is independent of part of its input and just applies to the rest of the argument.

The second construction is based on the existence of a one-way function with an almost-known preimage size. In other words, when an element in the domain is given, an estimate of the size of the preimage set is easily computable, with a polynomial uncertainty. An example of such a function is a regular function, which is a function such that every image of an n-bit input has the same number of preimages of length n. Another such function is decoding random linear codes. Subset sum is another example of such a function.

The details of the constructions are beyond the scope of this brief survey, as it needs other definitions and more explanations.

7.4.4 Rompel's Scheme

Rompel managed to construct a UOWHF from any one-way function [Rompel, 1990]. His construction is rather complicated and elaborate, and a detailed explanation is beyond the scope of this survey. However, the idea is to transform any one-way function into a UOWHF through a sequence of complicated procedures. First, the one-way function is transformed into another one-way function such that for most elements of the domain it is easy to find a collision, except for a fraction of them. From this, another one-way function is constructed such that for most of the elements it is hard to find a collision. Subsequently, a length increasing one-way function is constructed such that it is almost everywhere hard to find any collision. Finally this is turned into a UOWHF, which compresses the input such that it is difficult to find a collision.

7.5   Hard Bits and Pseudorandom Bit Generation

All the schemes presented in Section 7.4 are of theoretical importance, especially Rompel's scheme as it can be shown that one-way functions are necessary and sufficient for secure digital signatures. In other words, if there exists a one-way function, it is possible to construct a secure digital signature. However, they are rather impractical.

Zheng, Matsumoto and Imai showed that there is a duality between one-way hash functions and pseudorandom bit generators [Zheng *et al.*, 1990b]. A pseudorandom bit generator is a function that given a randomly chosen input, called a seed, outputs a longer string which cannot be efficiently distinguished from a truly random one. On the other hand, a one-way hash function generates a shorter string for a longer one given as the input. The output is called the hash value and it is computationally difficult to find a pair of strings that are compressed to the same hash value. They suggested a construction which is a dual of the Blum-Micali pseudorandom bit generator. For brevity, we call it the ZMI scheme. The goal of this construction is to provide a practical scheme, rather than reducing the assumptions on the complexity of the functions used. However, this construction has a limitation. It is impossible to compress more than $O(\log n)$ bits, due to their assumption which is the application of any one-way permutation with some known hard bits.

In this section, we present the notion of hard bits, which is the basis of the Blum-Micali pseudorandom bit generator and also the ZMI hash scheme, and the definition of the Blum-Micali pseudorandom bit generator. In the next section, we present a one-way permutation which we call *strong*, and we show how to apply it in order to achieve an efficient Blum-Micali pseudorandom bit generator, and an efficient hash scheme.

First an informal definition of hard bits. If a function f is one-way then given $f(x)$ the argument x must be unpredictable [Blum and Micali, 1984]. If every bit of the argument x were easily computable from $f(x)$, then f would not be a one-way function. Therefore, some specific bits of the argument are unpredictable, and we cannot guess them any better than by flipping a coin. We call these bits hard bits of f.

Definition 7.10 *Let $f : D \to R$ be a one-way function, where $R = \bigcup_n \Sigma^n$ and $D = \bigcup_n \Sigma^{l(n)}$. Let $i(n)$ be a function from N to N with $1 \leq i(n) \leq n$. If, for each probabilistic polynomial time algorithm M, for each polynomial Q and for all sufficiently large n,*

$$Prob\{M(f_n(x)) = x'_{i(n)}\} < \frac{1}{2} + \frac{1}{Q(n)}$$

where $x \in_r \Sigma^n$ and $x'_{i(n)}$ is the $i(n)$-th bit of an $x' \in \Sigma^n$ satisfying $f(x) = f(x')$, then the $i(n)$-th bit is a hard bit of f [Zheng et al., 1990a].

The definition of hard bits implies that under f^{-1} a hard bit depends on all bits of $f(x)$, where f^{-1} is a hard problem and does not run in polynomial time. Another intuitive description by Goldreich and Levin is that such bits *concentrate* the one-wayness of the function in a strong sense [Goldreich and Levin, 1989].

Note that the one-wayness of a function is relative to a specific model of computation with a specific amount of computing resources. On the other hand, the unpredictability or randomness of bits or strings are also relative to the specific model of computation with the specific amount of computing resources. In this chapter, we investigate n-bit one-way permutations and also the randomness of n-bit long strings, where the running time of an algorithm is a polynomial $Q(n)$ in the length of input. $Q(n)$ can be represented as being equal to $2^{\kappa(n)}$, where $\kappa(n)$ is of order $O(\log_2 n)$. κ may be used instead of Q whenever it is more convenient. For example, a *computing resource for k bits* is defined as follows.

Definition 7.11 *We say we have a* computing *resource for k bits if, given the output of a one-way function and $n - k$ bits of the input string, one can find the remaining k bits of the input string by exhaustive search.*

For the remainder of this chapter, we assume that the available computing resources are for less than k bits such that 2^{k+1} has a growth rate slightly greater than any polynomial in n.

Lemma 7.1 *The number of hard bits indicates the difficulty of inverting a one-way function if all the remaining bits are easily calculated.*

Proof : Assume that only a small number of bits of a function are hard bits and, when the output is given, we can obtain every remaining bit with a probability better than $\frac{1}{2} + \frac{1}{Q(n)}$ in polynomial time. A probabilistic algorithm M that first predicts the easy bits and then does an exhaustive search for finding hard bits can invert the function f in polynomial time with a probability at least $\frac{1}{Q(n)}$. For example, consider that a function has been proven to have only $\log_2(n)$ hard bits and all its remaining bits are easy to calculate. If $n = 512$ then only 9 bits are hard. If we have a computing resource for more than 9 bits, which we usually have, then given the output, the input can be obtained in polynomial time with a probability better than $\frac{1}{Q(n)}$. □

Hence, a one-way function which has $n - k - 1$ bits that are easy to predict, for example, appear directly in the output, should have $k + 1$ hard bits.

Lemma 7.2 *All the hard bits are independent of one another.*

Proof : We give a proof by contradiction. Assume that the i_1, i_2-th bits are hard bits that are dependent on each other and there is a probabilistic algorithm M that can calculate i_1-th bit given both $f(x)$ and i_2-th bit with a probability better than $\frac{1}{Q(n)}$. Then, we can construct a probabilistic algorithm M' for guessing the i_1-th bit.

Algorithm M':

1. Guess the i_2-th bit by flipping a coin (guess with probability 0.5).

2. Given the i_2-th bit and $f(x)$, run M and find the i_1-th bit.

then $Prob\{M'(f(x)) = x_{i_1}\} > \frac{1}{2} + \frac{1}{Q(n)}$; which is a contradiction. $\quad\square$

From the above lemma, the following corollary is drawn readily.

Corollary 7.1 *Let $f : D \to R$ be a one-way function, where $D = \bigcup_n \Sigma^n$ and $R = \bigcup_n \Sigma^{l(n)}$. Assume f has t hard bits, $t < n - k$, and $j < k$ of them and $f(x)$ are given, we cannot predict any of the remaining $t - j$ hard bits with a probability better than $\frac{1}{2} + \frac{1}{Q(n)}$.*

As we are going to describe the Blum-Micali pseudorandom bit generator, the formal definition of the *next bit test* is given below. The notion of next bit test was presented roughly in Chapter 4, where it was suggested that for a bit generator to be pseudorandom, it should pass the next bit test.

Definition 7.12 *Let l be a polynomial, and E be an ensemble with length $l(n)$. We say that E passes the next bit test if for each statistical test T, for each polynomial Q and for all sufficiently large n, the probability that on input the first i bits of a sequence x randomly selected according to E and $i < l(n)$, T outputs the $(i+1)$th bit of x is*

$$Prob\{T(x_1, \ldots, x_i) = x_{i+1}\} < \frac{1}{2} + \frac{1}{Q(n)}$$

where $x \in_E \Sigma^{l(n)}$.

The following theorem is derived from [Yao 82] and has been stated in [Alexi *et al.*, 1988], [Blum and Micali, 1984] and [Goldreich *et al.*, 1986] in a different form.

Theorem 7.3 *Let E be an polynomially samplable ensemble, the following statements are equivalent:*

- *E passes the next bit test.*

- *E is indistinguishable from the uniform ensemble U.*

In other words, the indistinguishability test is equivalent to the unpredictability test.

Corollary 7.2 *Assume that $f : D \to R$ is a one-way function, where $D = \bigcup_n \Sigma^n$ and $R = \bigcup_n \Sigma^{l(n)}$. Also assume that $i_1, i_2, \ldots, i_t$ are functions from N to N, with $1 \leq i_j(n) \leq n$ for each $1 \leq j \leq t$, $t < k$ and each i_j denotes a hard bit of f. Denote by E_n^1 and E_n^2 the probability distributions of the random variables $x_{i_t(n)} \ldots x_{i_2(n)} x_{i_1(n)} \parallel f(x)$ and $r_t \ldots r_2 r_1 \parallel f(x)$ respectively, where $x \in_r \Sigma^n$, $x_{i_j(n)}$ is the $i_j(n)$-th bit of x and $r_j \in_r \Sigma$. Let $E^1 = \{E_n^1 \mid n \in N\}$ and $E^2 = \{E_n^2 \mid n \in N\}$, then E^1 and E^2 are indistinguishable from each other.*

Proof :   From Corollary 7.1, it can be concluded that every string of $t < k$ hard bits passes the next bit test. This is equivalent to saying that given $f(x)$, any string of $t < k$ hard bits is indistinguishable from a string chosen uniformly at random from Σ^t, according to Theorem 7.1. $\Box$

In other words, given $f(x)$, any string of $t < k$ hard bits is indistinguishable from a random string. Such hard bits are called *simultaneous hard bits* of f. Note that the maximum number of simultaneous hard bits of any one-way function cannot be more than $n - k$.

The notion of hard-core predicates of functions was first discovered by Blum and Micali and was applied to construct pseudorandom bit generators (PBG). In the following PSB is defined.

Definition 7.13 *Let l be a polynomial with $l(n) > n$. A pseudorandom bit generator is a deterministic polynomial time function g that upon receiving a random n-bit input, extends it into a sequence of $l(n)$-bit pseudorandom bits $b_1, b_2, \ldots, b_{l(n)}$ as the output.*

In other words:

1. Each bit b_k is easy to compute.

2. The output bits are unpredictable, in other words the output string passes the next bit test, that is given the generator g and the first s output bits $b_1, \ldots, b_s$, but not the input string, it is not computationally feasible to predict the $(s+1)$th bit in the sequence

The following theorem describes the Blum-Micali PBG.

Theorem 7.4 *Let l be a polynomial with $l(n) > n$, and let f be a one-way permutation on $D = \bigcup_n \Sigma^n$ and let the $i(n)$-th bit be a proven hard bit of f. Let g_n be a function defined as follows:*

1. *Generate the sequence $f_n^{(1)}(x), f_n^{(2)}(x), \ldots, f_n^{(l(n))}(x)$, where $x \in \Sigma^n$.*

2. *From right to left (!), extract the i-th bit from each element in the above sequence and output that bit.*

so, $g_n(x) = b_{l(n)}(x) \ldots b_2(x)\, b_1(x)$ where $x \in \Sigma^n$ and $b_j(x) = $ (the i-th bit of $f_n^{(j)}(x)$). The $g = \{g_n \mid n \in N\}$ is a pseudorandom bit generator extending n-bit into $l(n)$-bit output strings.

If the $i_1(n)$, ..., $i_t(n)$-th bits are simultaneous hard bits of f, then the efficiency of g can be improved by defining the $b_j(x)$ to be a function which extracts all known simultaneous hard bits of $f^{(j)}(x)$. In [Alexi et al., 1988], it was proved that the $\log_2(n)$ least significant bits of the RSA and Rabin encryption functions are simultaneously hard. Hence, if we use the RSA or Rabin functions instead of the one-way permutation, with each iteration of the function we can extract $\log_2(n)$ bits. For example, if n is equal to 512 and we would like to produce a 512 bit pseudorandom string, we should iterate the one-way permutation for $\lceil \frac{l(n)}{\log_2(n)} \rceil = \lceil \frac{512}{\log_2(512)} \rceil = 57$ times. If a one-way permutation has more known hard bits, we can use it instead of the RSA or Rabin function and obtain better efficiency.

7.6   A Strong One-Way Permutation

In this section we construct a one-way permutation with maximum number of hard bits, which can be used for the construction of both the Blum-Micali pseudorandom bit generator and one-way hash functions. Before describing the construction some preliminary definitions are given. These definitions are from [Webster and Tavares, 1985].

Definition 7.14 *A transformation is called* complete *if each output bit depends on all input bits. In other words, the simplest Boolean expression for each output bit contains all the input bits.*

Definition 7.15 *If the inverse of a complete transformation is also complete, it is described as being* two-way complete. *In other words, each output bit depends on all the input bits and vice-versa.*

Lemma 7.3 *If a permutation is complete, then it is also two-way complete.*

Definition 7.16 *If the correlation between two binary variables is zero, they are called* independent variables.

We do not include the definition of correlation here, as it is not necessary. However, it is given in [Webster and Tavares, 1985] for the interested reader.

Definition 7.17 *Let v be a complete permutation and let all the output bits be pairwise independent. We call v a* perfect permutation.

Kam and Davida in [Kam and Davida, 1979] presented a method where an entire substitution-permutation network could be guaranteed to be complete, if all the substitution boxes used in the procedure were complete. DES is an example of a complete cryptographic transformation. Since DES is reversible and the inverse function (decryption) has the same structure as encryption, DES is a two-way complete transformation. Webster and Tavares showed that there is very little correlation between output variables of DES

[Webster and Tavares, 1985]. So, we can conclude that DES is an example of a perfect permutation in our definitions. Brown has used the known design criteria of DES to build an extended 128-bit DES and has shown that his scheme has similar cryptographic properties to DES [Brown, 1989]. Extending the DES structure for more bits, for example 512 bits, has the disadvantage that the running time is relatively high and comparable to that of a public key cryptosystem. For the following theorems, we use a two-way complete permutation such that only $k+1$ output bits are independent of the other bits and we call it a *$(k+1)$-bit perfect permutation*, which has much looser requirements than a perfect permutation.

Lemma 7.4 *Let f be an n-bit one-way permutation and V be the set of all n-bit permutations which are computable in polynomial time, then $m = f \circ v \circ f$ is also a one-way permutation, when $v \in_r V$.*

Proof : Both f and v are polynomial time computable permutations, so the result of their composition is a polynomial time computable permutation. It is also one-way as f is a one-way permutation. The probability that m would not be one-way, is equal to the probability of inverting f in polynomial time, and is less than $\frac{1}{Q(n)}$. □

By putting some conditions on v and f, we can make the one-way permutation m such that it would be a permutation with the desired properties.

Theorem 7.5 *Let $m : D \to D$ be a one-way permutation where $D = \bigcup_n \Sigma^n$ and $m = f \circ v \circ f$, where f is a one-way permutation and it has at least $k+1$ hard bits, and v is a $(k+1)$-bit perfect permutation where the positions of independent output bits correspond to the position of hard bits of f. For each probabilistic polynomial time algorithm M, for each polynomial Q and for all sufficiently large n,*

$$Prob\{M(m(x)) = x_i\} < \frac{1}{2} + \frac{1}{Q(n)}$$

where $x \in_r \Sigma^n$ and x_i is the i-th bit of the x, and $1 \le i \le n$. In other words, each bit of x is a hard bit of m.

Proof : We obtain our proof by contradiction. We show that given $m(x)$, if an algorithm could find x_i, it would be able to invert f. For simplicity of notation, we indicate the first one-way function by f_1 and the second one by f_2, so $m = f_2 \circ v \circ f_1$. Assume that M is an algorithm that given $m(x)$, can predict x_i with a probability bigger than $\frac{1}{2} + \frac{1}{Q(n)}$. In other words, x_i is not a hard bit of m. Two situations may arise:

1. When x_i is not a hard bit of f_1:

   Since the i-th bit is not a hard bit of f_1, given $f_1(x)$ there exists an algorithm M' that can find the i-th bit with a probability bigger than $\frac{1}{2} + \frac{1}{Q(n)}$. Without loss of generality, consider v to be an invertible permutation. Due to the two-way completeness property of v, all bits of $v \circ f_1(x)$ depend on all bits of $f_1(x)$ and vice-versa. So, to obtain $f_1(x)$, we need to know all the bits of $v \circ f_1(x)$. Since v is an invertible function in polynomial time, given $v \circ f_1(x)$, it is possible to find the i-th bit of x,

   $$\mathrm{Prob}\{M'(v \circ f_1(x)) = x_i\} > \frac{1}{2} + \frac{1}{Q'(n)}$$

   The probability equation simply says that we can predict x_i by tossing a coin with probability $1/2$ or estimating it given $v \circ f_1(x)$ with a probability better than $1/Q'(n)$. In other words,

   $$\mathrm{Prob}\{\text{estimating } x_i \mid v \circ f_1(x)\} > \frac{1}{Q'(n)}$$

   Without loss of generality, we assume that f_2 is a one-way permutation such that given a $f_2(y)$, we can guess $n - k - 1$ bits of y efficiently. Moreover, the $k + 1$ independent bits of v correspond to the hard bits of f_2, and knowing some other bits of $v \circ f_1(x)$ (that is, other than the independent output bits of v) and v, we cannot calculate all the bits of $v \circ f_1(x)$. In accordance with the assumption that the i-th bit is not a hard bit of m, the following also holds:

   $$\frac{1}{Q(n)} < \mathrm{Prob}\{\text{estimating } x_i \mid f_2 \circ v \circ f_1(x)\}$$
   $$= \mathrm{Prob}\{\text{estimating } x_i \mid v \circ f_1(x)\} \times$$
   $$\mathrm{Prob}\{\text{obtaining } v \circ f_1(x) \mid f_2 \circ v \circ f_1(x)\}$$

   Since the multiplication of two polynomial expressions is another polynomial expression, the following holds for some polynomial Q''.

   $$Prob\{\text{obtaining } v \circ f_1(x) \mid f_2 \circ v \circ f_1(x)\} > \frac{1}{Q''(n)}$$

This is equivalent to inverting f_2 and contradicts our assumption that f_2 is a one-way permutation.

2. When x_i is a hard bit of f_1:

by performing a procedure similar to the first case, it is obvious that the i-th bit should also be a hard bit of m.

□

Lemma 7.5 *Let $m = f_2 \circ v \circ f_1$ be a one-way permutation, where f_2 and v are defined as in Theorem 7.5, and let f_1 to be a one-way permutation such that given $t < n - k$ bits of x, no $\ell > k$ bits of $f(x)$ can be guessed with a probability better than $\frac{1}{2^k}$, then given $m(x)$ and any $t < n - k$ bits of x, $m(x)$ still cannot be inverted.*

Proof :   Since $t < n - k$ bits of x are known, the value of $f_1(x)$ can be guessed with a probability equal to $\frac{1}{2^{n-t}}$, where $n - t > k$. Hence, any bit of $v \circ f_1(x)$ cannot be estimated with a probability better than $\frac{1}{2^{n-t}} < \frac{1}{2^k}$, if v is a two-way complete permutation. Without loss of generality, assume that given $f_2 \circ v \circ f_1(x)$, $n-k-1$ bits of $v \circ f_1(x)$ can be guessed efficiently. Since the position of the hard bits of f_2 correspond to the positions of the independent bits of v, given $n - k - 1$ bits of $v \circ f_1(x)$, we cannot still estimate the $k + 1$ independent output bits of v with a probability better than $\frac{1}{2^{k+1}}$. The only possibility for reversing m is that the hard bits of f_2 and the $t < n - k$ bits of x be related to each other by some function such that revealing the t bits of x makes estimating the hard bits of f_2 probable. Such possibility has been excluded by assuming that f_1 is a one-way permutation such that given $t < n - k$ bits of x, no $\ell > k$ bits of $f(x)$ can be guessed with a probability better than $\frac{1}{2^k}$. Because, even if $v \circ f_1(x)$ and $f_1(x)$ are related to each other by a system of linear equations, knowing $n - k - 1$ bits of $v \circ f_1(x)$ and $\ell < k$ bits of the $f_1(x)$, the system of equations still cannot be solved.   □

Note that the conditions of Lemma 7.5 for f_1 only exclude one-way permutations that split into two or more parts, for example $f(x_1 \parallel x_2) = x_1 \parallel g(x_2)$. As the definition of hard bits implies that the hard bit affects all the output bits, if a one-way permutation with some hard bits had been used for f_1, then the above conditions would be satisfied without any further assumptions.

Lemma 7.6 *Let $m = f_2 \circ v \circ f_1$ be the one-way permutation defined in Theorem 7.5, then given $m(x)$ and any string of $t < n - k$ bits of x, m cannot be inverted.*

Lemma 7.6 suggests a simple construction for a one-way permutation m such that each bit of x is a hard bit of m and given any $t < n - k$ bits of x and $m(x)$, m cannot be inverted. We call such a permutation m a *strong one-way permutation* or simply a *strong permutation*. The following corollary can be drawn from Lemma 7.6.

Corollary 7.3 *Assume that $m : D \to D$ is a strong one-way permutation, where $D = \bigcup_n \Sigma^n$. Also assume that $i_1, i_2, \ldots, i_t$ are functions from N to N, with $1 \leq i_j(n) \leq n$ for each $1 \leq j \leq t$, $t < n - k$. Denote by E_n^1 and E_n^2 the probability distributions of the random variables $x_{i_t(n)} \ldots x_{i_2(n)} \, x_{i_1(n)} \parallel m(x)$ and $r_t \ldots r_2 \, r_1 \parallel m(x)$ respectively, where $x \in_r \Sigma^n$, $x_{i_j(n)}$ is the $i_j(n)$-th bit of x and $r_j \in_r \Sigma$. Let $E^1 = \{E_n^1 \mid n \in N\}$ and $E^2 = \{E_n^2 \mid n \in N\}$, then E^1 and E^2 are indistinguishable from each other.*

In other words, any string of $t < n - k$ bits of x is indistinguishable from a random string.

We can now construct an efficient Blum-Micali pseudorandom bit generator with the strong one-way permutation suggested in Theorem 7.5.

Theorem 7.6 *Let l be a polynomial with $l(n) > n$ and m be a strong one-way permutation. Let g be a function defined as follows:*

1. *Generate the sequence $m_n^{(1)}(x), m_n^{(2)}(x), \ldots, m_n^{(l(n))}(x)$, where $x \in \Sigma^n$.*

2. *From right to left, extract $n - k - 1$ bits from each element in the above sequence and output them.*

Then g is a pseudorandom bit generator extending n-bit input to $(n - k - 1)l(n)$-bit output strings.

Since we have a computing resource for k bits, the above scheme yields the maximum possible efficiency. If $k = 128$ (!) and n is 512, then with 2 iterations of m, or 4 iterations of f, we can extract 766 pseudorandom bits. This yields nearly 192 pseudorandom bits per iteration of f, which is 21 times more efficient than using the RSA or Rabin function with the scheme described in Theorem 7.4.

Note that since the output string is pseudorandom, we can also draw the following corollary.

Corollary 7.4 *The $n - k - 1$ extracted bits of each iteration is distributed uniformly and randomly in Σ^{n-k-1}.*

7.7   UOWHF Construction and PBG

Damgard in [Damgard, 1989] used pseudorandom bit generators for hash functions by extracting a small portion of the output string. Later, Zheng, Matsumoto and Imai revealed a duality between the construction of pseudorandom bit generators and one-way hash functions [Zheng *et al.*, 1990a]. We show that the construction presented in Theorem 7.6 for a PBG, can also be used for the construction of a UOWHF. Before showing this, we make some remarks about UOWHF's.

For universal one-way hash functions, there is no guarantee that it is not computationally feasible to find pairs of inputs that map onto the same output. However, there should not be too many such pairs. So, choosing x randomly, it should be unlikely that anyone can find an x' such that $h(x) = h(x')$ [Merkle, 1989b]. However, if we assume that h is random, that is, hashing is accomplished by looking up the correct value in a large table of random numbers, then it is possible to choose x in a non-random way since any method of choosing x that does not depend on h is random with respect to h.

Another problem with universal one-way hash functions is that repeated use weakens them. To deal with this problem, we can simply define a family of one-way hash functions with the property that each member h_i of the family is different from all other members, so any information about how to

break h_i will provide no help in breaking h_j for $i \neq j$ (see [Merkle, 1989b]).
If the system is designed so that every use of a weak one-way hash function is
parameterized by a different parameter, then the overall system security can
be kept high. The UOWHF that Naor and Yung constructed was based on
the application of a one-way permutation and a strongly universal$_2$ family of
hash functions. In the next subsection, we show how to construct a UOWHF
by applying the strong one-way permutation presented in Section 7.6.

7.7.1   UOWHF Based on the Strong One-way Permutation

The following theorem describes the construction of the UOWHF.

Theorem 7.7 *Assume that* $m : D \rightarrow D$ *is a strong one-way permutation,*
where $D = \bigcup_n \Sigma^n$, *and* $\text{chop}_1 : \Sigma^n \rightarrow \Sigma^{n-1}$ *simply chops the last bit, then*
$h = \text{chop}_1 \circ m$ *is a universal one-way hash function.*

Proof :   We obtain our proof by contradiction. Assume that there is a
probabilistic algorithm F that can find a collision, then we show that we can
make an algorithm that can invert m. Suppose that we first choose an x at
random, then run m on x to get $m(x)$, we then obtain $h(x) = \text{chop}_1(m(x))$.
There is only one element that can collide with $m(x)$ under chop_1. This
element differs from $m(x)$ in one bit. Let us write this element as $m(y)$. If
a collision finder can find a y which collides with x under h with probability
greater than $\frac{1}{Q(n)}$, it can obtain y from $m(y)$ with the same probability. This
contradicts our assumption that m is a one-way permutation.   □

Lemma 7.7 *If we define* $1\text{chop} : \Sigma^n \rightarrow \Sigma^{n-1}$ *to chop one bit and the position*
of the chopped bit is given in the description of the function and can be any
bit, then $h = 1\text{chop}(m(x))$ *is also a universal one-way hash function.*

Proof Sketch: The problem of finding a collision for h, defined in Lemma
7.7, can be reformulated to finding x, y and $x \neq y$, such that $m(x)$ and $m(y)$
match at all bits except at the one specified in the definition of the *1chop*

function. By repeating a procedure similar to that for the proof of Theorem 7.7 the claim of this lemma can be shown to be true. □

Since according to Corollary 7.3 and Corollary 7.4 the output of m is distributed uniformly and randomly in Σ^n, then to find y by exhaustive search, we need to perform 2^{n-1} operations on average. If this much computation is greater than 2^k, then it is not feasible to find the collision.

If we chop t bits of $m(x)$, then there are $(2^t - 1)$ elements which collide with x under h. If these elements are distributed randomly among 2^n elements, then we need to do 2^{n-t} search operations to find a collision for x. Since our computational resource can do at most 2^k search operations then t should be less than $n - k$.

Corollary 7.5 *Let* $\text{chop}_t : \Sigma^n \to \Sigma^{n-t}$ *chop the* t *last bits and let* $t < n - k$, *then* $h = \text{chop}_t \circ m$ *is a universal one-way hash function.*

Note that the scheme described in the above corollary increases the efficiency of the hash function scheme, so for hashing long messages, we need to do less iterations. We can also generalize the above scheme by introducing *tchop*, a function which chops t bits of the output. In this case, we need $(n - k - 1) \log_2 n$ bits to specify the positions of the chopped bits.

7.7.2 Parameterization

Since the hash function presented in Corollary 7.5 is a universal one-way hash function, we should parameterize it to make it secure for implementation in a practical scenario. The parameterization can be done in two different ways:

1. We can parameterize h by selecting v from a family of $(k+1)$-bit perfect permutations. Then $H = \{h = \text{chop}_t \circ f \circ v \circ f \mid v \in V_n\}$ where V_n is the $(k + 1)$-bit perfect permutation family and chop_t chops the t last bits.

2. We can parameterize h by selecting the function for the compressing procedure from a family of hash functions. We may choose a family of *chop* functions. In this case, the number of bits required to specify a

member of the family is at most equal to $(n-k-1)\log_2 n$. Alternatively, we could choose a family of t to 1 strongly universal hash function as proposed in [Naor and Yung, 1989].

7.7.3 Compressing Arbitrary Length Messages

One of the main desirable properties of hash functions is that they should be applicable to an argument of any size. Damgard suggested a design principle in [Damgard, 1989] based on fixed size collision free hash functions. Another method appeared in [Zheng *et al.*, 1990a] and is the dual of the Blum-Micali pseudorandom bit generator. Let us call it the ZMI method. We show that using the strong one-way permutation proposed in Theorem 7.5, these two methods actually yield one scheme for hashing long messages.

Damgard's method: Let $l(n)$ be a polynomial with $l(n) > n$, let f be a collision free one-way hash function $f : \Sigma^{n+t} \to \Sigma^n$ and let $\alpha \in_r \Sigma^n$. Split an $l(n)$-bit message x into t-bit blocks and let the blocks be denoted by $x_1, x_2, \ldots, x_{\frac{l(n)}{t}}$. If

$$y_0 = \alpha$$
$$\vdots$$
$$y_{i+1} = f(y_i \parallel x_{i+1})$$

then $h(x) = y_{\frac{l(n)}{t}}$ is the hash value of the long message x.

ZMI method: Let f be a one-way permutation $f : \Sigma^{n+t} \to \Sigma^{n+t}$ and let $I(n) = (i_1, i_2, \ldots, i_t)$ denote the known simultaneously hard bits of f. Let $x = x_t \ldots x_2 x_1 \in \Sigma^t$ and $b \in \Sigma^n$. Define $\text{ins}_{I(n)}(b, x)$ be a function inserting bits of x in the i_1-th, $\ldots$, i_t-th bits of b, that is:

$$\text{ins}_{I(n)}(b, x) = b_n \ldots b_{i_t} x_t b_{i_t-1} \ldots b_{i_1} x_1 b_{i_1-1} \ldots b_2 b_1$$

Let $z \in \Sigma^{n+t}$ and denote by $\text{drop}_{I(n)}(z)$ a function dropping the i_1-th, $\ldots$, i_t-th bits of z. Let l be a polynomial with $l(n) > n$ and let $\alpha \in \Sigma^n$. Split an $l(n)$-bit message x into t-bit blocks denoted by $x_1, x_2, \ldots, x_{\frac{l(n)}{t}}$, where $x_i \in \Sigma^t$ for each $1 \le i \le \frac{l(n)}{t}$. Let h be the function from $\Sigma^{l(n)}$ to Σ^n defined by:

$$y_0 = \alpha$$

$$y_1 = \text{drop}_{I(n)}(f(\text{ins}_{I(n)}(y_0, x_{\underset{t}{\iota(n)}})))$$

$$\vdots$$

$$y_i = \text{drop}_{I(n)}(f(\text{ins}_{I(n)}(y_{i-1}, x_{\underset{t}{\iota(n)}-i+1})))$$

then $h(x) = y_{\underset{t}{\iota(n)}} = \text{drop}_{I(n)}(f(\text{ins}_{I(n)}(y_{\underset{t}{\iota(n)}-1}, x_1)))$. In the original
ZMI scheme $h(x) = f(\text{ins}_{I(n)}(y_{\underset{t}{\iota(n)}-1}, x_1))$. If we use the strong one-way
permutation m in the ZMI scheme for f, since the t least significant bits
are simultaneously hard bits, then the $\text{drop}_{I(n)}$ function performs identically
to the chop_t function defined in Corollary 7.5. So, $\text{drop}_{I(n)}(f(x))$ in the
ZMI method would be identical to $\text{chop}_t(m(x))$ of Corollary 7.5, which is a
universal one-way hash function from Σ^{n+t} to Σ^n. On the other hand, when
the t last bits of a function are simultaneously hard bits, then $\text{ins}_{I(n)}(y_0, x_{\underset{t}{\iota(n)}})$
would yield the same result as $(y_0 \parallel x_{\underset{t}{\iota(n)}})$. So using the strong one-way
permutation with the ZMI scheme would yield the same result as using the
one-way hash function proposed in Corollary 7.5 with Damgard's method,
when the message blocks are fed in a similar order.

7.8   A Single construction for UOWHF and PBG

Each iteration of the pseudorandom bit generator presented in Theorem 7.6
is identical to the hash function presented in Corollary 7.5. Assume that we
have a computational resource for at most $k=63$ bits. For the construction
of the PBG of Theorem 7.6, an algorithm should extract at most $n - k - 1$
bits, and throw away at least $k+1$ bits on each iteration. On the other hand,
for the construction of the one-way hash function according to Corollary 7.5,
we may chop at most $n - k - 1$ bits, and leave $k + 1$ bits as the hash value.
If we choose $k < t < n - k$, for example for $n = 512$ we choose $64 \leq t \leq 448$,
then the algorithm can be used both for pseudorandom bit generation and
universal one-way hashing.

7.9   Conclusions and Extensions

1. We constructed a strong permutation with a $(k+1)$-bit perfect permutation, namely a complete permutation whose $k+1$ output bits are independent. A $(k+1)$-bit perfect function can be constructed easily as follows:

$$v(x) = c(x) \oplus PBG_{k+1}(x)$$

where $x \in \Sigma^n$, $c(x)$ is a complete permutation and $PBG_{k+1}(x)$ denotes $k+1$ output bits of a pseudorandom bit generator, for the seed x. However, we constructed a UOWHF and also an efficient pseudorandom bit generator with the strong permutation. This confirms Naor and Yung's conjecture that if pseudorandom bit generators exist then UOWHF's exist [Naor and Yung, 1989].

2. For the construction of the strong permutation we assumed that the position of $k+1$ hard bits of the one-way function f corresponds to that of the $k+1$ independent bits of v. The following generalization can easily be shown to be true.

   If v is a perfect permutation then $m = f \circ v \circ f$ is a strong one-way permutation, where f is any one-way permutation.

   In other words, there is no need to know the exact positions of the hard bits of f. As we mentioned earlier, the running time of a perfect permutation based on a DES structure for large enough n, for example $n=512$, is rather big.

   A reasonable question is whether we can apply some simpler mathematical functions, such as $y = (a_i x)^3 \bmod m$, and/or $y = (ax + b) \bmod m$, or a composition of such functions for v. In Chapter 8, we investigate how multiple compositions of polynomials of $y = px + q$ in $GF(2^n)$ with a one-way permutation can be employed to construct a strong one-way permutation.

Chapter 8

How to Construct a Family of Strong One-way Permutations

8.1   Introduction

Much effort has been spent to identify the hard bits of some specific number theoretic one-way functions. In [Alexi *et al.*, 1988] it is shown that the $O(\log n)$ least siginficant bits of the RSA and Rabin encryption functions are individually hard, and that those $O(\log n)$ bits are also simultaneously hard. In addition, it is shown in [Long and Wigderson, 1988] that the exponentiation function, that is $f(x) = g^x(\bmod P)$, where P is a prime and g is a generator of Z_P^*, also has $O(\log n)$ hard bits. Both these works take advantage of complicated techniques based on number theoretic approaches.

A breakthrough in this area is due to Goldreich and Levin [Goldreich and Levin, 1989] who have shown how to build a hard-core predicate for all one-way functions. They have extended the construction to show that $O(\log n)$ pseudorandom bits can be extracted from any one-way function. Their result cannot be improved without imposing additional assumptions on the one-way function [Goldreich and Levin, 1989], leaving the problem of constructing a function with $O(n)$ simultaneous hard bits open. However, if a one-way function is proven to have a higher degree of security, then a greater number of pseudorandom bits could be extracted using the same method.

In [Blum and Goldwasser, 1985], [Goldwasser and Micali, 1984] a construction for a probabilistic encryption function was presented, for which all the bits of the presented one-way function are simultaneously hard. This construction is based on the composition of hard bits from many one-way functions.

Another significant work in this area is due to Scherift and Shamir. They have shown in [Scherift and Shamir, 1990] that half of the input bits of an exponentiation function (modulo Blum integers), that is $f(x) = g^x (\mathrm{mod}\ N)$ where N is a Blum integer, are simultaneously hard and almost all bits are individually hard to evaluate. This work also takes advantage of complicated techniques. As a result, exponentiation modulo a Blum integer is as yet the only natural function with $O(n)$ proven simultaneous hard bits.

In Chapter 7, an n-bit one-way permutation such that each input bit is individually hard and any $t < n - O(\log n)$ input bits are simultaneously hard is called a strong one-way permutation. In this chapter, we show how to construct a family of strong one-way permutations, such that all input bits of a permutation are hard and any $t < n - O(\log n)$ input bits are indistinguishable from a random string. In contrast with [Goldwasser and Micali, 1984], which composes the hard bits of many one-way functions, we compose one-way permutations to get a strong one-way permutation. Two practical schemes are proposed. Both schemes take advantage of the family of polynomials in a Galois field. The first scheme is based on the existence of a one-way permutation and is constructed with $O(\frac{n}{\log n})$ fold composition of a one-way permutation, and a randomly chosen element of the family of polynomials in a Galois field. The second scheme is based on the existence of a hiding one-way permutation, and is constructed with a three layer structure applying a hiding one-way permutation, a randomly chosen element of the family of polynomials in a Galois field and any one-way permutation.

Section 8.2 gives some preliminary comments on hard bits. In Section 8.3, after investigating some properties of polynomials in a Galois field, the proposed constructions are presented.

The results of this chapter have appeared in [Sadeghiyan *et al.*, 1991].

8.2 Preliminary Comments

We gave the formal definition of one-way functions in Chapter 7. However, informally speaking, one-way functions are those which are easy to compute but difficult to invert. It is clear that the one-way property of a function is relative to a specific amount of computing resources in a specific model of computation. In this chapter, we assume that we have a computing resource for at most $2^{k(n)}$ operations, where $k(n) = O(\log n)$. We also assume $k^+(n)$ is a function with a growth rate slightly more than $k(n)$ such that $n - k^+ > k$. As an example, consider $k^+(n) = O(\log^{1.01}(n))$.

If a function f is one-way then given $f(x)$ the argument x must be unpredictable. If every bit of the argument x were easily computable from $f(x)$, then f would not be a one-way function. Hard bits are some specific bits of the argument which are unpredictable and cannot be guessed with a probability better than by flipping a coin [Blum and Micali, 1984], [Zheng *et al.*, 1990a]. There may exist some one-way functions which do not have any hard bits. However, we require that any one-way function should have more than $k(n) = O(\log n)$ bits which are unpredictable, though they may be biased.

If a one-way permutation f had $t = n - k^+$ (the maximum possible) known simultaneous hard bits, it could be used in the Blum-Micali pseudorandom bit generator scheme to obtain the maximum efficiency for the generator, where the maximum number of bits per iteration of f can be extracted. We called such a one-way permutation a strong one-way permutation or simply a strong permutation[1].

In the next section, we present two schemes for the construction of strong one-way permutations, where we take advantage of polynomials over the Galois field $GF(2^n)$. In the following, some properties of these polynomials which are of interest to us are investigated. However, first we consider the notion of strongly universal$_2$ hash functions, presented by Carter and Wegman in [Wegman and Carter, 1981].

---

[1]Note the term '*strongly* one-way permutation' has been used in [Goldreich *et al.*, 1988] with a different meaning, however, as the term *strong one-way permutation* conveys our desired meaning we used it with the new definition.

Definition 8.1 *Suppose G is a set of functions and each element of G is a function from A to B. G is strongly* universal$_2$ *if given any two distinct elements a_1, a_2 of A and any two elements b_1, b_2 of B, then $\frac{(\#G)}{(\#B)^2}$ functions take a_1 to b_1 and a_2 to b_2.*

In other words, the values of $g(x)$ and $g(y)$ are independently and uniformly distributed in B for every $x, y \in A$, when $g \in G$ is chosen uniformly at random. Strongly universal$_2$ sets of functions can be created using polynomials over finite fields. As the simplest example consider $G = \{g \mid g(x) = px + q; p, q \in GF(2^n)\}$ in the finite field $GF(2^n)$.

A Few Observations

In Chapter 7, we defined complete permutations as permutations where each output bit depended on all input bits. In other words, the Boolean expression for each output bit contained all the input bits. Since the operation in the Galois field $GF(2^n)$ is done modulo an irreducible polynomial, the resulting permutation is such that the Boolean expression for each output bit contains all the input bits. This is due to the properties of operations in Galois fields.

Example: We investigate the case for $GF(2^3)$. Here, x is a string of three bits, x_2, x_1, x_0, and represents the polynomial $x_2 z^2 + x_1 z + x_0$. Let p represent $p_2 z^2 + p_1 z + p_0$ and q be $q_2 z^2 + q_1 z + q_0$. There are two irreducible polynomials in $GF(2^3)$, $z^3 = z + 1$ and $z^3 = z^2 + 1$. When the operation is done modulo $z^3 = z + 1$:

$$
\begin{aligned}
g(x) = px + q = \quad & (q_2 + \underline{p_2 x_0 + p_1 x_1 + p_0 x_2} + p_2 x_2)z^2 \\
+ \ & (q_1 + \underline{p_1 x_0 + p_0 x_1 + p_2 x_2} + p_2 x_1 + p_1 x_2)z \\
+ \ & (q_0 + \underline{p_0 x_0 + p_2 x_1 + p_1 x_2})
\end{aligned}
$$

and when the operation is done modulo $z^3 = z^2 + 1$:

$$
\begin{aligned}
g(x) = px + q = \quad & (q_2 + \underline{p_2 x_0 + p_1 x_1 + p_0 x_2} + p_2 x_2 + p_2 x_1)z^2 \\
+ \ & (q_1 + \underline{p_1 x_0 + p_0 x_1 + p_2 x_2})z \\
+ \ & (q_0 + \underline{p_0 x_0 + p_2 x_1 + p_1 x_2} + p_2 x_2)
\end{aligned}
$$

Both irreducible polynomial have produced some common terms, which are functions of all the input bits, and some other terms in the coefficients of

$g(x)$. This would happen if the operation was performed in any $GF(2^n)$. So, it is clear that polynomials in $GF(2^n)$ result in a complete permutation. Notice that when p, q are chosen at random, for every x, y, the outputs are uniformly and independently distributed.

When we are operating in $GF(2)$ the multiplication is equivalent to the 'AND' operation and addition is equivalent to the 'XOR' operation. So each coefficient in $g(x)$ is the inner product of x with a different string obtained from p. Note that, if we represent x and $g(x)$ as two vectors, they are related to each other by a system of n linear equations. In the above example when the irreducible polynomial is $z^3 = z + 1$ the above equations can be represented as

$$g(x) = px + q = \left(\begin{bmatrix} p_0 + p_2 & p_1 & p_2 \\ p_1 + p_2 & p_0 + p_2 & p_1 \\ p_1 & p_2 & p_0 \end{bmatrix} \begin{bmatrix} x_2 \\ x_1 \\ x_0 \end{bmatrix} + \begin{bmatrix} q_2 \\ q_1 \\ q_0 \end{bmatrix} \right) \begin{bmatrix} z^2 & z^1 & z^0 \end{bmatrix}$$

If we had applied polynomials of higher degree, such as $g(x) = \alpha x^2 + \beta x + \gamma$, a similar result would have been obtained. For the remainder of this chapter, we use the simplest case $g(x) = px + q$, although all the following lemmas and theorems are also true when g is of a higher degree. Notice that the above relation can be stated in a vector representation as:

$$\mathbf{g(x) = px + q}$$

Moreover, $\mathbf{p}$ and $\mathbf{q}$ can be modified in a way such that $\mathbf{g}$ and $\mathbf{x}$ become related to each other through a Toeplitz matrix. A Toeplitz matrix is a matrix M such that $M_{i,j} = M_{i+1,j+1}$ for all i, j. For the above example, we may write

$$\mathbf{g} = \begin{bmatrix} p_0 + p_2 + r_2 & p_1 & p_2 \\ p_1 + p_2 + r_1 & p_0 + p_2 + r_2 & p_1 \\ p_1 + r_0 & p_1 + p_2 + r_1 & p_0 + p_2 + r_2 \end{bmatrix} \begin{bmatrix} x_2 \\ x_1 \\ x_0 \end{bmatrix} + \begin{bmatrix} q_2' \\ q_1' \\ q_0' \end{bmatrix}$$

where $\mathbf{r} = \begin{bmatrix} r_2 \\ r_1 \\ r_0 \end{bmatrix}$ is a randomly chosen vector, and $q_2' = q_2 + r_2 x_2$ and $q_1' = q_1 + r_1 x_2 + r_2 x_1$ and $q_0' = q_0 + r_0 x_2 + (p_1 + r_1)x_1 + (p_2 + r_2)x_0$.

Lemma 8.1 *If $g(x) = px + q$, where $p, q \in GF(2^n)$ are chosen randomly, and $n - k^+$ bits of the $g(x)$ are known (or its k^+ bits are unknown), then the probability of guessing each bit of x is equal to $\frac{1}{2^{k^+}}$.*

Proof :   Assume that there is an algorithm L which when given p, q and some bits of $g(x)$ lists all possible values of x. Since g is a permutation, if one bit of $g(x)$ is given, L will list all possible values of x which will be 2^{n-1} elements on average. In general, given i bits of $g(x)$, L lists 2^{n-i} possible values of x. If $n - k^+$ bits of $g(x)$ are given, L will list 2^{k^+} possible elements for x. One can guess the correct value of x with a probability of $\frac{1}{2^{k^+}}$. Since g is a complete permutation and p and q are chosen randomly, the overall probability of guessing any bit of x is equal to the probability of guessing the value of x.   □

Note that, for some specific values of p, q some bits of x could be guessed efficiently, but when we consider the probability of guessing any bit of x over all values of p and q, it is equal to $\frac{1}{2^{k^+}}$.

8.3   Strong One Way Permutations

In this section we propose two schemes for the construction of strong one-way permutations. The first construction is given in Theorem 8.1: the building block of that construction is $f \circ g$. The following lemma investigates $f \circ g$.

Lemma 8.2 *Let $m : D \to D$ be a one-way permutation where $D = \bigcup_n \Sigma^n$ and $m = f \circ g$, where f is a one-way permutation, and where $g = px + q$ with $p, q \in_r GF(2^n)$. Also assume that $i_1, i_2, \ldots, i_k$ are functions from N to N, with $1 \le i_j(n) \le n$ for each $1 \le j \le k$. Denote by E_n^1 and E_n^2 the probability distributions of the random variables $x_{i_k(n)} \cdots x_{i_2(n)} \, x_{i_1(n)} \parallel m(x)$ and $r_k \ldots r_2 \, r_1 \parallel m(x)$ respectively, where $x \in_r \Sigma^n$, $x_{i_j(n)}$ is the $i_j(n)$-th bit of x and $r_j \in_r \Sigma$. Let $E^1 = \{E_n^1 \mid n \in N\}$ and $E^2 = \{E_n^2 \mid n \in N\}$, then E^1 and E^2 are indistinguishable from each other.*

In other words, given $m(x)$, the probability of distinguishing any k-bit string of x from a random string is less than $\frac{1}{2^k}$, where $k = O(\log n)$ when the probability is calculated over all values of p, q. Note that Lemma 8.2 virtually says that, *given $f(x')$ and $p', q' \in_r GF(2^n)$ it is hard to guess any $O(\log n)$ bits of x,* where $f(x') = f \circ g(x) = m(x)$. As $x = p'x' + q'$ is the inverse of $g(x) = x' = px + q$, $p, q \in_r GF(2^n)$. So x is actually the concatenation of the inner products of x' with n different strings obtained from p'.

Proof : Goldreich and Levin showed in [Goldreich and Levin, 1989] that:

given $f(x')$ and p', where f is any one-way function and $\mid p' \mid = \mid x' \mid$ and p' is an arbitrary string, the inner product of x' and p' is a hard-core predicate of f, and cannot be guessed with a probability better than $\frac{1}{2} + \frac{1}{Q(|x'|)} = \frac{1}{2} + \frac{1}{Q(n)}$ for each probabilistic polynomial time algorithm and for each Q.

They also extended their result and showed that $O(\log n)$ hard bits can be obtained from any one-way function, where the simultaneous hard bits are the inner product of $O(\log n)$ different n-bit strings with x'. According to [Goldreich and Levin, 1989], the set of strings may also form a Toeplitz matrix. As was mentioned earlier, the matrix which relates x to $g(x)$ can be rearranged into a Toeplitz matrix, so the same sort of proof that has been given in [Goldreich and Levin, 1989] could be presented here to show that any k bits of x are indistinguishable from a random string when $m(x)$ is given. As the method that Goldreich and Levin used to proved their claim is rather involved and complicated, we avoid repeating it here. However, a simple and informal justification can be given as follows.

Without loss of generality, assume that f is a one-way permutation which acts on k^+ bits and keeps the other bits unchanged. So, given $m(x) = f \circ g(x)$, it is hard to guess k^+ bits of $g(x)$, but $n - k^+$ bits of $g(x)$ can be guessed efficiently. As was proved earlier, x and $g(x)$ are related to each other with a system of n equations with n variables. When $n - k^+$ bits of $g(x)$ are known, the system of equations can be reduced to a system in k^+ variables. However, if any k^+-bit string of x is given, the system of equations can be solved and the values of the k^+ unknown bits of $g(x)$ would be revealed. Since there is no algorithm which can invert f with a probability better than $\frac{1}{2^{k^+}}$, no bit of x can be guessed with an overall probability better than $\frac{1}{2} + \frac{1}{2^{k^+}}$. Moreover, any probabilistic algorithm M that could distinguish any $t \leq k$ bit string of x from a random string with a probability better than $\frac{1}{2^{k^+}}$ would be able to invert f with an overall probability better than $\frac{2^t}{2^{k^+}}$ which contradicts our assumption that f is a one-way permutation. $\qquad\square$

The result of Lemma 8.2 can be compared with the results of Vazirani and Vazirani in [Vazirani and Vazirani, 1984], where it is shown that the XOR of any non-empty subset of hard bits is also hard to guess. Altogether, it can be concluded that all bits of x are individually hard, and any $k = O(\log n)$

bits of x are simultaneously hard bits of $f \circ g$ and cannot be distinguished from a random string with a probability of success better than $\frac{1}{2^k}$, when the probability is computed over all values of p and q.

8.3.1   A Scheme for the Construction of Strong Permutations

With the Goldreich-Levin method, only $O(\log n)$ pseudorandom bits can be extracted from any one-way function. This number of pseudorandom bits cannot be improved without additional assumptions on the complexity of the one-way function used. The reason this is true is that a one-way function which cannot be inverted with a probability better than $\frac{1}{Q(n)}$, may act only on $\log Q(n)$ of the bits of x and leave the rest unchanged [Goldreich and Levin, 1989]. In Lemma 8.2, we constructed a one-way permutation $f \circ g$ such that any k input bits cannot be distinguished from a random string. If we apply $f \circ g$ as a one-way permutation in the Blum-Micali pseudorandom bit generator, any k bits can be extracted per iteration of $f \circ g$. We take advantage of such a one-way permutation to construct a family of strong permutations.

In the following, we suggest two schemes to obtain strong permutations and present the theorems behind them. The first scheme is based on the s fold composition of the $f \circ g$.

Theorem 8.1 *Let $m : D \to D$ be a one-way permutation where $D = \bigcup_n \Sigma^n$ and $m = (f \circ g)^s = \underbrace{(f \circ g) \circ \ldots \circ (f \circ g)}_{s \text{ times}}$, where f is a one-way permutation, and $s = O(\frac{n}{\log n})$, and $g = px + q$ where $p, q \in_r GF(2^n)$. Then m is a strong one-way permutation.*

Proof :   First, we show that $(f \circ g)^2 = f \circ g \circ f \circ g$ has $2k$ hard bits. Let us denote the first k bits string of x to be $x_{\leftarrow 1}$, its second k bits to be $x_{\leftarrow 2}$ and so on, and consider $y = f \circ g(x)$. Given $(f \circ g)^2(x)$, the string $x_{\leftarrow 2} \parallel y_{\leftarrow 1}$ is indistinguishable from a random string (this is true because the concatenation of hard bits from each iteration in Blum-Micali pseudorandom bit generator is indistinguishable from a random string and according to

Lemma 8.2 any k bit of $f \circ g$ is indistinguishable from a random string). As $x_{\leftarrow 2} \parallel y_{\leftarrow 1}$ forms a $2k$-bit string, $x \oplus y$ has $2k$ bits which cannot be guessed efficiently. Since any k bits of $f \circ g$ are indistinguishable from a random string, any $2k$-bit string of $x \oplus y$ is also indistinguishable from a random string. So, for each probabilistic polynomial time algorithm M

$$\text{Prob}\{M[(f \circ g)^2(x)] = x \oplus y\} < \frac{1}{2^{2k+}}$$

In the following, we will show that the above relation implies that:

$$\text{Prob}\{M[(f \circ g)^2(x)] = x\} < \frac{1}{2^{2k+}}$$

To justify this claim, by contradiction assume that there is a probabilistic algorithm M' that can compute x with probability better than $\frac{1}{2^{2k}}$. By applying M' in another probabilistic algorithm M'', it can be shown that the value of $x \oplus y$ can be computed with a probability better than $\frac{1}{2^{2k}}$. M'' first runs M' on $(f \circ g)^2(x)$ to get the value of x. Then M'' runs $f \circ g$ on x to find the value of y. M'' gives $x \oplus y$ as its output. If the value of x is correct, the value of y would be correct with probability 1. Hence, M'' outputs the correct value for $x \oplus y$ with a probability better than $\frac{1}{2^{2k}}$. This contradicts our assumption that $x \oplus y$ cannot be guessed with probability better than $\frac{1}{2^{2k}}$. So, it can be concluded that for each probabilistic polynomial time algorithm M

$$\text{Prob}\{M[(f \circ g)^2(x)] = x\} < \frac{1}{2^{2k+}}$$

In this way, a one-way permutation is obtained which is more complex than f, without putting any condition on f. As the number of pseudorandom bits extracted from a one-way function (or the number of simultaneous hard bits) with the Goldreich-Levin method is bounded by the complexity or the 'security parameter' (Goldreich and Levin's term) of the one-way function, then $2k$ simultaneous hard bits can be extracted from $(f \circ g)^2$. This can be done by choosing a random $2k \times n$ Toeplitz matrix, and multiplying it by x. Note that, since p and q are chosen randomly and independently, the matrix which relates $g(x)$ to x can be arranged in a Toeplitz matrix form. In addition, the matrix can be arranged in such a way that for any predetermined $2k$ bits of x, the corresponding rows form a Toeplitz matrix. Hence, given $(f \circ g)^2(x)$, any $2k$ bits of x cannot be distinguished from a random string. This completes the proof that two fold iteration of $f \circ g$ produces a one-way function such that every $2k$ input bits are indistinguishable from a random string.

Using induction and performing a proof similar to that above, it can be shown that $(f \circ g)^i(x)$ has a complexity or security factor such that ik random bits can be extracted from it. In addition, in a similar way it can be shown that each input bit is individually a hard bit and any ik input bits are simultaneous hard bits. Therefore, to obtain a one-way permutation with any $n - k$ input bits simultaneously hard, it is enough to construct an $\frac{n}{k}$ fold iteration of $f \circ g$. With $k = O(\log n)$, a construction of $O\left(\frac{n}{\log n}\right)$ fold iteration of $f \circ g$ is needed, which would be performed in polynomial time anyhow.   □

8.3.2   A Three-layer Construction for Strong Permutations

In Lemma 8.2, we constructed a one-way permutation $f \circ g$ such that any k input bits cannot be distinguished from a random string. If there exists another transformation (permutation) h (such that given any t bit string of its input x, where $t < n - k$, it will be difficult to guess any k bits of its output $h(x)$), then we can apply this function before g and get $f \circ g \circ h(x)$ as a one-way permutation. The probability of distinguishing any $t < n - k$ bit string of x from random strings is less than $\frac{1}{2^{k+}}$, when its output is given. This result is proved in the next subsection. Since h would be able to hide any k bits of its output, we call it a hiding permutation.

First, we introduce a definition for hiding permutations.

Definition 8.2 *Assume that $h : D \to D$ is a permutation. Also assume that $i_1, \ldots, i_t$ and $j_1, \ldots, j_k$ are functions from N to N, where $1 \le i_l(n), j_l(n) \le n$ for each $1 \le l \le n$. We call h a* hiding permutation, *if for each probabilistic polynomial time algorithm M, for each $t < n - k$ and for each polynomial Q and for all sufficiently large n,*

$$| \, Prob\{M(x_{i_t}, \ldots, x_{i_1} \, \| \, y_{j_n}, \ldots, y_{j_{k+}}) = y_{j_k}, \ldots, y_{j_1}\} - \frac{1}{2^k} \, | < \frac{1}{Q(n)}$$

where $x \in_r \Sigma^n$ and x_i denotes i-th bit of x, and y_j denotes j-th bit of $h(x)$.

The following theorem shows how to make a strong permutation from a hiding permutation.

Theorem 8.2 *If h is a hiding permutation and $g = px + q$, where $p, q \in_r$ $GF(2^n)$, and f is any one-way permutation, then $m = f \circ g \circ h$ is a strong one-way permutation.*

Proof Sketch: To prove that m is a strong one-way permutation we need to show that any bit of x is hard to guess, and any $n - k^+$ bits of x are simultaneously hard, given only $m(x)$. Assume that, by contradiction, a probabilistic polynomial time algorithm M could guess x_i, given $m(x)$. Let $x' = h(x)$. Since any x_i is a function of some bits of x', and according to Lemma 8.2, any input bit of $f \circ g$ is individually a hard bit, then any algorithm which can guess x_i can guess a hard bit of $f \circ g$. This is contradictory to the assumption that f is a one-way permutation, since computing any hard bit of $f \circ g$ is equivalent to reversing f. Hence, every x_i is a hard bit of m. Moreover, having $t < n - k$ bits of x does not reveal any k bits of $h(x)$, since h is a hiding permutation. Then, having $t < n - k$ bits of x would not help in inverting m, given $m(x) = f \circ g \circ h(x)$. So, for each polynomial Q and for large enough n, any probabilistic polynomial time algorithm M cannot distinguish any string of $t < n - k$ bits of x from a random string with a probability better than $\frac{1}{Q(n)}$, when $m(x)$ is given. $\qquad\square$

A method for hiding x is based on the application of a one-way permutation which acts on all bits, and serves as a hiding permutation due to the following lemma.

Lemma 8.3 *Any one-way permutation h which is complete, is a hiding permutation.*

Proof : We obtain a proof by contradiction. Assume that a one-way permutation which is complete is not a hiding permutation. Then, there is a probabilistic polynomial time algorithm M that can obtain $y_{j_k}, \ldots, y_{j_1}$, given $x_{i_t}, \ldots, x_{i_1}$, with the available computing resources. On the other hand, since $n - t > k$ bits of x are not given, then the k bits of the output obtained do not depend on at least $n - t - k$ bits of the input. This is equivalent to saying that

h is formed from two functions, say h_1, h_2, with $h(x) = h_1(x_{i_{t+k}}, \ldots, x_{i_1}) \parallel h_2(.)$. Obviously h_1 is not a function of x, which contradicts our assumption that h is a complete permutation.   □

Thus, a concrete example of the construction of a family of strong one way permutations, based on using a complete one-way permutation as the hiding permutation, is $m(x) = f \circ g \circ h(x)$, where f is any one way permutation, $g = px + q$ where $p, q \in_r GF(2^n)$, and h is a complete one-way permutation which acts on all its bits.

8.4   Conclusions

There are many functions that are considered to be one-way, so if someone knows the value of $f(x)$, he can find the value of x for less than a fraction $\frac{1}{Q(n)}$ of x's. This does not necessarily mean that any bit of x cannot be guessed efficiently. On the other hand, it is shown that $O(\log n)$ bits of the RSA and the Rabin encryption schemes are hard to guess [Alexi *et al.*, 1988]. Also, it was shown that $O(\log n)$ bits of the exponentiation function are hard to guess [Long and Wigderson, 1988]. This does not mean that the remaining bits are easy to guess, but only that we do not yet have any proof about the remaining bits. Recently it was shown that $\frac{n}{2}$ bits of the exponentiation function are simultaneously hard to guess, when the operation is done modulo a Blum integer [Scherift and Shamir, 1990].

In this chapter, we showed how to make a family of strong one-way permutations, such that whenever a member is chosen uniformly at random, we get a one-way permutation such that all its input bits are hard and any $t < n - k$ bit string of input bits is indistinguishable from a random string, with a high probability. Two schemes for this purpose were suggested. The first scheme is based on a one-way permutation. The second scheme relies on the existence of a hiding permutation. An open problem is to show that a one-way permutation is complete, or cannot be split into two parts. We also took advantage of the simplest family of polynomials in a Galois field and showed that it is also a family of complete permutations: it had already been shown that it is a family of strongly *universal*$_2$ functions. The proposed schemes for the construction of a family of strong one-way permutations can be shown to work with families of polynomials of higher degree in Galois fields as well, where such polynomials form a family of strongly universal$_n$ functions. As it was shown in Chapter 7, a strong one-way permutation is an effective tool for the construction of efficient pseudorandom bit generators and universal one-way hash functions.

Chapter 9

Conclusions

This book has reviewed some existing cryptographic hash functions together with methods of attacks on them, and has developed some principles for the design of such functions. The results of the review and the development of design principles may now be summarized.

9.1 Summary

Chapter 1 introduced the aim of the book and provided a background for the theory and practice of secure hash schemes. The necessity for information authentication in computer environment was explained and cryptographic primitives to provide security mechanisms were presented. Particular emphasis was given to the notion of digital signature and digital signatures with RSA encryption systems were introduced. Signature-hashing schemes were described as an improvement over a digital signature scheme with RSA, since some algebraic properties of RSA can be exploited to produce forged digital signatures. Signature-hashing schemes provide not only better security but also other desired properties such as efficiency. There were many proposals for hash schemes, but, with a few exceptions, their security was left as an open problem. The proposals were later analyzed and most of them were found to be insecure. As the security question of many cryptographic schemes and services reduces to the question of existence of a secure hash scheme, the aim of the book was to develop some design principles for the construction of secure hash functions.

Chapter 2 surveyed the area of hash functions in cryptography and provided an overview of the different schemes proposed. Requirements for secure hash functions were described, and different types of hash functions with their specifications were also given. Hash functions were divided into strong or collision-free hash functions and weak or universal one-way hash functions, based on their degree of security. They were also divided into MAC, where a private key takes part in the scheme, or MDC, otherwise. From the structural point of view, they were divided according to whether they applied a block cipher as the underlying one-way function, in which case they were called block-cipher-based hash functions, or whether they used any other function which was easy to compute but considered difficult to invert. Finally, principles for designing a hash function which hashes messages of several block sizes, given a hash function which only hashes one message block, were presented and two methods, namely, the serial method and the parallel method were described.

Chapter 3 described different methods of attack on hash functions. The birthday attack can be launched against any hash scheme. The probability of success depends on the length of the hash value. For a 64-bit hash value, gathering 2^{33} hash values and messages increases the chance of finding two messages having the same hash value to more than 63 %. This happens if the hash scheme performs a random mapping; otherwise it would be possible to take advantage of the non-random behaviour of the hash scheme to find two colliding messages with fewer operations. The other attacks depend on the structure of the hashing scheme. The meet-in-the-middle attack can be launched against hashing schemes which employ block chaining in their structure. The correcting-the-last-block attack can be launched against hash schemes based on CBC mode of DES, or on modular arithmetics. Some other attacks take advantage of weaknesses, such as weak keys or some weak structural algebraic properties, in the algorithm. The differential cryptanalysis attack takes advantage of the non-uniform distribution properties of the hash scheme.

Chapter 4 examined the notion of pseudorandomness, defined basic ideas such as indistinguishability, and described recent developments in this area. It also explained the relationship between this notion and the design of block-cipher-based hash functions. Based on the idea of indistinguishability, pseudorandom bit generators, pseudorandom function generators and pseudorandom permutation generators were defined. The construction of pseudorandom permutation generators from pseudorandom function generators is attributed to Luby and Rackoff. Their construction employs three rounds of DES-like permutations and three pseudorandom function generators. When a permutation generator is pseudorandom, it is secure against chosen plaintext attack. Luby and Rackoff considered this as a justification for the application of DES-like permutations in the structure of DES. Different trials for reducing the number of pseudorandom functions together with their weaknesses were then described. The notion of super-pseudorandomness was presented. If a cryptosystem is super-pseudorandom, it is secure against a chosen plaintext/ciphertext attack. It is possible to make a super-pseudorandom permutation generator with four rounds of DES-like permutations and four pseudorandom function generators. This has the implication that it is possible to achieve better security by adding to the number of rounds in DES-like cryptosystems.

The meet-in-the-middle attack against a block-cipher-based hash scheme is a super-distinguishing circuit for the underlying block cipher. If the block cipher is secure against chosen plaintext/ciphertext attack, the meet-in-the-middle attack cannot successfully be applied against the corresponding block-cipher-based hash scheme. Hence we are interested in developing a structure which can be used for the construction of cipher systems secure against chosen plaintext/ciphertext attacks, so that it can be used for the construction of block-cipher-based hashing algorithms.

Chapter 5 examined the construction of super-pseudorandom permutations and presented necessary and sufficient conditions for achieving such permutations. The conclusion drawn from the chapter was that a composition of DES-like permutations is super-pseudorandom if and only if the two internal structures, that is, one without the first round and one without the last round, are not only pseudorandom but also independent permutations. An important corollary of this result is that it is possible to construct a super-pseudorandom permutation generator with four DES-like permutations and only two pseudorandom function generators. This structure employs one pseudorandom function generator for the first and second rounds and another one for the third and the fourth rounds. It was also shown that a four-round DES-like structure with a single pseudorandom function generator is not super-pseudorandom, although it may be pseudorandom. The other contribution of this chapter is the investigation of the conditions for super-pseudorandomness of generalized DES-like permutations. The major result of this part is that it is possible to construct super-pseudorandom permutations with k^2 rounds of type-1 Feistel type permutations, where k is the number of branches of the structure.

Chapter 6 described a structure with a single pseudorandom function generator and six rounds of DES-like permutations which was super-pseudorandom. First it was shown that the composition of Luby and Rackoff permutation generators is also pseudorandom. However, it was shown it is possible to achieve a perfect randomizer by composing two Luby and Rackoff permutation generators by applying permutations instead of functions for the intermediate layers. A structure with four random functions was presented which was also shown to be perfect. Then it was shown that the same structure, with only two pseudorandom function generators, is super-pseudorandom. Finally, it was shown that replacing one of the function generators with a two-fold composition of the other one does not affect the super-pseudorandomness of the permutation generator. In the same manner that the result of Luby and Rackoff may be considered to be a justification for the application of DES-like permutations in the structure of DES and DES-like block ciphers, this structure can be recommended for the design of block ciphers with better security so that they can be used in block-cipher-based hash schemes.

As the other type hash functions are based on one-way functions other than block ciphers, Chapters 7 and 8 assume that a one-way permutation is given, and examine the construction of such hash functions. One-way functions are functions that are easy to compute but difficult to invert. Chapter 7 presented a construction for one-way hash functions and pseudorandom bit generators. This chapter first reviewed some complexity-theoretic constructions for hash functions, on the assumption that a one-way function exists. These constructions, although important from a theoretical point of

view, are less than practical. As hash functions can be considered as duals of pseudorandom bit generators, a practical scheme called the ZMI hash
scheme, which is dual to the Blum-Micali pseudorandom bit generator, was
based on the existence of a one-way permutation. This scheme, although
practical, suffers from low efficiency. To improve it, a one-way permutation,
which was called a strong one-way permutation, was constructed based on a
three layer structure. The first and the last layer of the construction was a
one-way permutation with some known hard bits and the intermediate layer
was a complete permutation with some independent output bits, where the
positions of the independent bits correspond to the positions of the hard bits
of the one-way permutation. Applying this strong permutation, an efficient
Blum-Micali pseudorandom bit generator and an efficient ZMI hash scheme
could be constructed. A concluding remark of this chapter was that, by applying the strong one-way permutation, the ZMI hash scheme is reduced to
the general design of Damgard for compressing long messages.

Chapter 8 examined a family of one-way functions that provides a practical proposal for the construction of strong one-way permutations. This
family has the property that, when a member is drawn randomly, every bit
of the input is a hard bit and every string of $n - O(\log n)$ bits of input is
simultaneously hard. This construction consists of $\frac{n}{O(\log n)}$ times composition
of a one-way permutation with a family of strongly universal$_2$ permutations.
With this structure, we are able to transform any one-way permutation into
an efficient hash function by applying the construction proposed in Chapter
7. However, it is also possible to reduce the number of layers to three by
applying a one-way function called a hiding permutation.

9.2   Limitations and Assumptions of the Results

This book has been primarily concerned with developing some principles
for the design of hash functions. In Chapter 2, we roughly divided hash
functions into two groups. The first group consisted of those schemes which
employ block ciphers in their structures. The design of block ciphers is
based on the theoretical work of Shannon, who proposed a substitution -

permutation network for the provision of confusion and diffusion of the bits in the construction of cryptographic algorithms. This led to the development of block ciphers such as DES, FEAL and LOKI.

Lai and Massey showed that, for a block-cipher-based hash scheme, any attack on its block cipher implies an attack of the same type on the hash scheme with the same complexity [Lai and Massey, 1992]. The meet-in-the-middle attack against a block-cipher-based hash scheme is a super-distinguishing circuit for the underlying block cipher. If the block cipher is super-pseudorandom, the meet-in-the-middle attack cannot successfully be applied against the corresponding block-cipher-based hash scheme, when our computational resources allow us to work on a polynomial (in the length of input) variation of the message. In other words, if the underlying block cipher acts like a random permutation in the face of chosen plaintext/ciphertext attack, or is secure against chosen plaintext/ciphertext attack, then the block-cipher-based hash scheme is secure against meet-in-the-middle attack. Hence, it can be recommended to use block ciphers secure against chosen plaintext/ciphertext attack in block-cipher-based hash schemes. It can be recommended to apply structures for block-cipher-based hash schemes and the block ciphers to be used in such schemes which make super-pseudorandomness achievable.

Unfortunately, the known block ciphers are at most claimed to be secure against chosen plaintext attacks, and none of them claim to be secure against a chosen plaintext/ciphertext attack. Most of the known block ciphers take advantage of DES-like structures. We restricted our investigation to the DES-like structure, and we tried to improve it so that it would be possible to construct a block cipher secure against chosen plaintext/ciphertext attack. Hence, we were interested in developing a structure which could be used for the construction of cipher systems secure against chosen plaintext/ciphertext attacks, so that it could be used for the construction of *block-cipher-based* hashing algorithms. We investigated necessary and sufficient conditions to achieve super-pseudorandomness for DES-like structures. We also showed that k^2 rounds of type-1 Feistel type permutations would yield a super-pseudorandom permutation, where k is the number of branches of the structure. We showed that $\psi(g,g,f,f)$ and $\psi(g,1,f,g,1,f)$ are super-pseudorandom. We also managed to construct a structure with a single pseudorandom function generator. The result was: $\psi(f^2,1,f,f^2,1,f)$, which is a six-round DES-like structure with a single pseudorandom function generator f. In the layers 1 and 4 of this construction, f is used directly, the layers 2 and 5 just XOR one branch to the other one, and in the layers 3 and 6 a two-fold composition of f is used. Note that $\psi(f,f,f,f,f,f)$ is not even pseudorandom, but $\psi(f^2,1,f,f^2,1,f)$ is super-pseudorandom, while no additional computation with respect to $\psi(f,f,f,f,f,f)$ is needed. So $\psi(f^2,1,f,f^2,1,f)$ can be adopted for the structure of block ciphers to be used for block-cipher-based hash schemes, as it allows us to achieve super-pseudorandomness and it requires only a single pseudorandom function generator.

The above results are based on the existence of a pseudorandom function generator. A method for the construction of pseudorandom function generators was given by Goldreich, Goldwasser and Micali, and was based on the existence of a pseudorandom bit generator. However, the existence of pseudorandom bit generators depends on the existence of one-way functions, and it is not yet known whether a one-way function exists as it depends on whether $P \neq NP$. In practice, designers attempt to achieve good S boxes, instead of designing pseudorandom function generators. Hence, although the proposed structure is sound, we have not yet constructed such a block cipher, as we still need to solve other problems such as the design of good S boxes and key scheduling or both.

The second group of hash schemes consisted of those schemes which are based on one-way functions other than block ciphers. One-way functions are those which are easy to compute but difficult to invert. As many different proposals for such schemes exists, we approached the problem from a theoretical point of view. We developed some generalized constructions for hash functions from one-way permutations. In fact, we assumed that a one-way permutation existed and we build a generalized construction for hash functions. First we noticed that the ZMI hash scheme can be improved if a one-way permutation with a greater number of simultaneously hard bits was incorporated in the scheme. Strong one-way permutations were defined and a three-layer construction, assuming the existence of a one-way permutation with $k + 1$ known hard bits and a $(k + 1)$-bit perfect permutation, was proposed for its construction. It was also shown, it is possible to achieve a strong one-way permutation with $\frac{n}{O(\log n)}$ times composition of any one-way permutation with a family of strongly universal$_2$ permutations. This structure can be adopted to transform a one-way permutation into a strong one-way permutation, and hence to obtain a hash function by applying the constructions proposed in Section 7.7.2. The proposed schemes can be applied with any one-way permutation. A limitation is the necessity for proving that a permutation in fact is one-way.

For functions such as RSA or exponentiation modulo a prime, the size of the arguments should be big enough so that it would be infeasible to invert them. For example, in the case of RSA, an argument length bigger than 512 bits is recommended. A drawback is that, applying such functions as the underlying one-way permutation makes the hash scheme rather time-consuming.

However, it was also possible to reduce the number of layers to three, by assuming the existence of a one-way function which was called a hiding permutation.

9.3 Prospects for Further Research

There have been many proposals for hash schemes, and some of them have been in use for a while. However, with time, most of them have been broken. One of the major reasons they were broken is advancement in technology. Once 2^{33} operations were far beyond the capability of computing resources; nowadays this is within reach. At that time hash functions producing a 64-bit result were reasonable designs, while today they are not. Today, if a cryptographic mechanism requires 2^{64} operations to be broken, it is considered secure, but who knows what the capabilities of future computers will be ?! Perhaps in the near future this number of operations would be quite accessible.

The unsuccessful efforts of many researchers who spent their time trying to design practical hash schemes suggest that we should work on some guidelines or principles for the design of hash functions, instead of proposing just another function which temporarily would seem secure. These designs should be adaptable to the capabilities of new technology. In this book, we developed some of these principles in such a way that we are not restricted to some specific number of bits. At the end of each chapter, we included related open problems, so we do not repeat them here. However, there is one open problem that should be mentioned here. As we saw earlier, to achieve a secure hash scheme the fundamental requirement is a one-way function which is easy to compute. If this problem can be solved, or at least a function found such that there is a considerable difference between the time necessary for its computation and its inversion, it would strengthen all the efforts made for the design of hash functions.

Bibliography

[Akl, 1983] S.G. Akl. On the Security of Compressed Encoding. In *Advances in Cryptology - CRYPTO '83*, pages 209–230. Plenum Publishing Corporation, 1983.

[Alexi *et al.*, 1988] W. Alexi, B. Chor, O. Goldreich, and C. P. Schnorr. RSA and Rabin Functions: Certain Parts Are As Hard As the Whole. *SIAM Journal on Computing*, 17(2):194–209, 1988.

[Banieqbal and Hilditch, 1990] B. Banieqbal and S. Hilditch. The Random Matrix Hashing Algorithm. Technical Report UMCS-90-9-1, Department of Computer Science, University of Manchester, 1990.

[Baritaud and Gilbert, 1992] T. Baritaud and H. Gilbert. F.F.T. Hashing is not Collision-free. In *Abstracts of Eurocrypt '92*, pages 31–40, 1992.

[Berson, 1992] T. A. Berson. Differential Cryptanalysis Mod 2^{32} with Applications to MD5. In *Abstracts of Eurocrypt '92*, pages 67–76, 1992.

[Biham and Shamir, 1990] E. Biham and A. Shamir. Differential Cryptanalysis of DES-like Cryptosystems. In *Abstracts of CRYPTO '90*, pages 1–19, 1990.

[Biham and Shamir, 1991a] E. Biham and A. Shamir. Differential Cryptanalysis of FEAL and N-hash. In *Abstracts of EUROCRYPT '91*, pages 1–8, 1991.

[Biham and Shamir, 1991b] E. Biham and A. Shamir. Differential Cryptanalysis of Snefru, Khafre, REDOC-II, LOKI and Lucifer. In *Abstracts of CRYPTO '91*, pages 4.1–4.7, 1991.

[Blum and Goldwasser, 1985] M. Blum and S. Goldwasser. An Efficient Probabilistic Public-Key Encryption Scheme Which Hides All Partial Information. In *Advances in Cryptology - CRYPTO '84*, volume 196 of *Lecture Notes in Computer Science*, pages 289–299. Springer-Verlag, 1985.

[Blum and Micali, 1984] M. Blum and S. Micali. How to Generate Cryptographically Strong Sequences of Pseudo-Random Bits. *SIAM Journal on Computing*, 13(4):850–864, 1984.

[Brown, 1989] L. Brown. A Proposed Design for an Extended DES. In *Computer Security in the Age of Information*. North-Holland, 1989. Proceedings of the Fifth IFIP International Conference on computer Security, IFIP/Sec '88.

[Brown, 1991] L. Brown. *Analysis of the DES and the Design of LOKI Encryption Scheme*. PhD thesis, University College, University of New South Wales, April 1991.

[Camion and Patarin, 1991] P. Camion and J. Patarin. The Knapsack Hash Function proposed at Crypto '89 can be broken, 1991. *In Abstracts of EUROCRYPT '91*.

[Carter and Wegman, 1979] J. L. Carter and M. N. Wegman. Universal Classes of Hash Functions. *Journal of Computer and System Sciences*, 18:143–154, 1979.

[Charnes and Pieprzyk, 1992] C. Charnes and J. Pieprzyk. Linear nonequivalence versus nonlinearity. In *Abstracts of AUSCRYPT'92*, Gold Coast, December 1992, pages 4.4-4.11.

[Coppersmith, 1985] D. Coppersmith. Another Birthday Attack. In *Advances in Cryptology - CRYPTO '85*, Lecture Notes in Computer Science, pages 14–17. Springer-Verlag, 1985.

[Coppersmith, 1989] D. Coppersmith. Analysis of ISO/CCITT Document X.509 Annex D, 1989.

[Daemen *et al.*, 1991a] J. Daemen, A. Bosselaers, R. Govaerts, and J. Vandewalle. Collisions for Schnorr's Hash Function Fft-Hash Presented at Crypto '91, 1991. presented at the rump session of ASIACRYPT '91.

[Daemen *et al.*, 1991b] J. Daemen, R. Govaerts, and J. Vandewalle. A Framework for the Design of One-way Hash Finctions Including Cryptanalysis of Damgård's One-way Function Based on a Cellular Automaton, 1991. in Abstracts of ASIACRYPT '91.

[Damgard, 1987] I. B. Damgard. Collision Free Hash Functions and Public Key Signature Schemes. In *Advances in Cryptology - EUROCRYPT '87*, volume 304 of *Lecture Notes in Computer Science*, pages 203–216. Springer-Verlag, 1987.

[Damgard, 1989] I. B. Damgard. A Design Principle for Hash Functions. In *Advances in Cryptology - CRYPTO '89*, volume 435 of *Lecture Notes in Computer Science*, pages 416–427. Springer-Verlag, 1989.

[Davida, 1982] G. I. Davida. Chosen signature cryptanalysis of the RSA public key cryptosystem. Technical Report TR-CS-82-2, Dept. of Electrical Engineering and Computer Science, Univ. of Wisconsin, 1982.

[Davies and Price, 1980] D. W. Davies and W. L. Price. The Application of Digital Signatures Based on Public Key Cryptosystems. In *Proceedings of the Fifth International Conference on Computer Communication*, pages 525–530, 1980.

[Davies and Price, 1984] D. W. Davies and W. L. Price. Digital Signature - an update. In *Proceedings of the Seventh International Conference on Computer Communication*, pages 845–849, 1984.

[Davies, 1983] D. W. Davies. Applying the RSA Digital Signature to Electronic Mail, 1983.

[De Jonge and Chaum, 1986] W. De Jonge and D. Chaum. Some Variations on RSA Signatures and their Security. In *Advances in Cryptology - CRYPTO '86*, volume 263 of *Lecture Notes in Computer Science*, pages 49–59. Springer-Verlag, 1986.

[De Santis and Yung, 1990] A. De Santis and M. Yung. On the Design of Provably-Secure Cryptographic Hash Functions. In *Abstracts of EUROCRYPT '90*, pages 377–397, 1990.

[den Boer and Bosselaers, 1991] B. den Boer and A. Bosselaers. An Attack on the Last Two Rounds of MD4. In *Advances in Cryptology - CRYPTO '91*, volume 576 of *Lecture Notes in Computer Science*, pages 194–203. Springer-Verlag, 1991.

[Denning, 1984] D. E. Denning. Digital Signatures with RSA and Other Public-Key Cryptosystems. *Communications of the ACM*, 27(4):388–392, 1984.

[DES, 1983] Data encryption algorithm - modes of operation, 1983. ANSI X3.106-1983.

[DES, 1985] Data encryption algorithm, 1985. AS 2805.5.

[Diffie and Hellman, 1976] W. Diffie and M. Hellman. New Directions in Cryptography. *IEEE Transactions on Information Theory*, IT-22(6):644–654, 1976.

[Garey and Johnson, 1979] M. Garey and D.S. Johnson. Computers and Intractibility: A guide to the Theory of NP-completeness. *W.H. Freeman and Co.*, San Francisco, 1979.

[Girault et al., 1988] M. Girault, R. Cohen, and M. Campana. A Generalized Birthday Attack. In *Advances in Cryptology - EUROCRYPT '88*, volume 330 of *Lecture Notes in Computer Science*, pages 129–156. Springer-Verlag, 1988.

[Girault, 1987] M. Girault. Hash-Functions Using Modulo-N Operations. In *Advances in Cryptology - EUROCRYPT '87*, volume 304 of *Lecture Notes in Computer Science*, pages 218–226. Springer-Verlag, 1987.

[Goldreich and Levin, 1989] O. Goldreich and L. A. Levin. A Hard-Core Predicate for all One-way Functions. In *the 21st ACM Symposium on Theory of Computing*, pages 25–32, 1989.

[Goldreich et al., 1986] O. Goldreich, S. Goldwasser, and S. Micali. How to Construct Random Functions. *Journal of the ACM*, 33(4):792–807, 1986.

[Goldreich et al., 1988] O. Goldreich, H. Krawczyk, and M. Luby. On the Existence of Pseudorandom Generators. In *Proceedings of the 29th IEEE Symposium on the Foundations of Computer Science*, pages 12–24, 1988.

[Goldwasser and Micali, 1984] S. Goldwasser and S. Micali. Probabilistic Encryption. *Journal of Computer and System Sciences*, 28:270–299, 1984.

[Goldwasser *et al.*, 1988] S. Goldwasser, S. Micali, and R. L. Rivest. A Digital Signature Scheme Secure against Adaptive Chosen-Message Attacks. *SIAM Journal on Computing*, 17(2):281–308, 1988.

[Harari, 1984] S. Harari. Nonlinear Non Commutative Functions for Data Integrity. In *Advances in Cryptology - EUROCRYPT '84*, Lecture Notes in Computer Science, pages 25–32. Springer-Verlag, 1984.

[Hellman *et al.*, 1976] M. Hellman, R. Merkle, R. Schroeppel, and L. Washington. Results of an Initial Attempt to Cryptanalyse the NBS Data Encryption Standard. Technical report, Information Systems Lab., Department of Electrical Engineering, Stanford University, 1976.

[Impagliazzo *et al.*, 1989] R. Impagliazzo, L. A. Levin, and M. Luby. Pseudorandom generation from one-way functions. In *the 21st ACM Symposium on Theory of Computing*, pages 12–24, 1989.

[Jueneman *et al.*, 1985] R. R. Jueneman, S. M. Matyas, and C. H. Meyer. Message Authentication. *IEEE Communication Magazine*, 23(9):29–40, 1985.

[Jueneman, 1982] R. R. Jueneman. Analysis of Certain Aspects of Output Feedback Mode. In *Advances in Cryptology - CRYPTO '82*, pages 99–127. Plenum Press, 1982.

[Jueneman, 1986] R. R. Jueneman. A High Speed Manipulation Detection Code. In *Advances in Cryptology - CRYPTO '86*, volume 263 of *Lecture Notes in Computer Science*, pages 327–347. Springer-Verlag, 1986.

[Jueneman, 1987] R. R. Jueneman. Electronic Document Authentication. *IEEE Network Magazine*, 1(2):17–23, 1987.

[Kam and Davida, 1979] J. B. Kam and G. I. Davida. Structured Design of Substitution-Permutation Encryption Networks. *IEEE Transactions on Computers*, 28(10):747–753, 1979.

[Lai and Massey, 1992] X. Lai and J. L. Massey. Hash Functions Based on Block Ciphers. In *Abstracts of Eurocrypt '92*, pages 53–66, 1992.

[Levin, 1987] L. A. Levin. One-Way Functions and Pseudorandom Generators. *Combinatorica*, 7(4):357–363, 1987.

[Long and Wigderson, 1988] D. L. Long and A. Wigderson. The Discrete
Logarithm Hides $O(\log n)$ Bits. *SIAM Journal on Computing*, 17(2):363–
372, 1988.

[Luby and Rackoff, 1988] M. Luby and C. Rackoff. How to Construct Pseudorandom Permutations from Pseudorandom Functions. *SIAM Journal
on Computing*, 17(2):373–386, 1988.

[Meijer and Akl, 1982] H. Meijer and S. Akl. Digital Signature Schemes.
Cryptologia, 6:329–338, 1982.

[Merkle, 1978] R. C. Merkle. Secure Communications over Insecure Channels. *Communications of the ACM*, 21(4):294–299, 1978.

[Merkle, 1979] R. C. Merkle. *Secrecy, Authentication, and Public Key Systems*. UMI Research Press, 1979.

[Merkle, 1989a] R. C. Merkle. A Certified Digital Signature. In *Advances
in Cryptology - CRYPTO '89*, volume 435 of *Lecture Notes in Computer
Science*, pages 218–238. Springer-Verlag, 1989.

[Merkle, 1989b] R. C. Merkle. One Way Hash Functions and DES. In *Advances in Cryptology - CRYPTO '89*, volume 435 of *Lecture Notes in Computer Science*, pages 428–446. Springer-Verlag, 1989.

[Merkle, 1990a] R. C. Merkle. Break of 2-pass snefru - reward for 4-pass
snefru, 1990. Newsgroups: sci.crypt.

[Merkle, 1990b] R. C. Merkle. Fast Software Encryption Functions. In *Abstracts of CRYPTO '90*, pages 457–473, 1990.

[Merkle, 1990c] R. C. Merkle. A Fast Software One-way Hash Function.
Journal of Cryptology, 3:43–58, 1990.

[Merkle and Hellman 1978] R. C. Merkle and M. E. Hellman. Hiding information and signatures in trapdoor knapsacks. *IEEE Trans. Inform.
Theory*, volume IT-24(5), September 1978, pages 525-530.

[Meyer and Matyas, 1982] C. H. Meyer and S. M. Matyas. *Cryptography: a New Dimension in Data Security*. Wiley & Sons, 1982.

[Mitchell and Walker, 1988] C. Mitchell and M. Walker. Solutions to the multidestination secure electronic mail problem. *Computers & Security*, 7:483–488, 1988.

[Mitchell, 1989] C. Mitchell. Multi-destination secure electronic mail. *The Computer Journal*, 32:13–15, 1989.

[Miyaguchi *et al.*, 1989] S. Miyaguchi, M. Iwata, and K. Ohta. New 128-bit Hash Function. In *Proceedings of 4th International Joint Workshop on Computer and Communications*, pages 279–288, 1989.

[Miyaguchi *et al.*, 1990] S. Miyaguchi, K. Ohta, and M. Iwata. Confirmation that Some Hash Functions Are Not Collision Free. In *Abstracts of EUROCRYPT '90*, pages 293–308, 1990.

[Montolivo and Wolfowicz, 1987] E. Montolivo and W. Wolfowicz. Digital Signature: an open problem. In *System Security 87*, pages 173–183, Pinner, Middx, UK, 1987. Online Publications.

[Moore, 1988] J. H. Moore. Protocol Failures in Cryptosystems. *Proceedings of the IEEE*, 76(5):594–601, 1988.

[Naor and Yung, 1989] M. Naor and M. Yung. Universal One-way Hash Functions and their Cryptographic Applications. In *the 21st ACM Symposium on Theory of Computing*, pages 33–43, 1989.

[Nishimura and Sibuya, 1990] K. Nishimura and M. Sibuya. Probability To Meet in the Middle. *Journal of Cryptology*, 2:13–22, 1990.

[Ohnishi, 1988] Y. Ohnishi. A study on data security. Master's thesis, Tohoku University, 1988. in Japanese.

[Ohta and Koyama, 1990] K. Ohta and K. Koyama. Meet-in-the-Middle Attack on Digital Signature Schemes. In *Abstracts of Auscrypt '90*, pages 110–121, 1990.

[Patarin, 1992] J. Patarin. How to Construct Pseudorandom and Super Pseudorandom Permutations from One Single Pseudorandom Function. In *Abstracts of Eurocrypt '92*, pages 235–245, 1992.

[Pieprzyk and Sadeghiyan, 1991] J. Pieprzyk and B. Sadeghiyan. Optimal Perfect Randomizers. In *Abstracts of ASIACRYPT '91*, pages 130–135, 1991.

[Pieprzyk and Zhang, 1990] J. Pieprzyk and X. Zhang. Permutation Generators of Alternating Groups. In *Advances in Cryptology - AUSCRYPT '90*, volume 453 of *Lecture Notes in Computer Science*, pages 237–244. Springer-Verlag, 1990.

[Pieprzyk, 1990] J. Pieprzyk. Theory of Pseudorandomness and its Application to Cryptography. Technical Report CS 90/15, University College, The University of New South Wales, 1990.

[Pieprzyk, 1991] J. Pieprzyk. How to Construct Pseudorandom Permutations from Single Pseudorandom Functions. In *Advances in Cryptology - EUROCRYPT '90*, volume 473 of *Lecture Notes in Computer Science*, pages 140–150. Springer-Verlag, 1991.

[Preneel *et al.*,] B. Preneel, R. Govaerts, and J. Vandewalle. Collision resistant hash functions based on blockciphers. submitted to CRYPTO '91.

[Preneel *et al.*, 1992] B. Preneel, R. Govaerts, and J. Vandewalle. Cryptographically Secure Hash Functions: an Overview, 1992.

[Quisquater and Delescaille, 1989a] J. J. Quisquater and J. P. Delescaille. How Easy is Collision Search? Application to DES. In *Abstracts of EUROCRYPT '89*, volume 434 of *Lecture Notes in Computer Science*, pages 429–434. Springer-Verlag, 1989.

[Quisquater and Delescaille, 1989b] J. J. Quisquater and J. P. Delescaille. How Easy is Collision Search? New results and applications to DES. In *Advances in Cryptology - CRYPTO '89*, volume 435 of *Lecture Notes in Computer Science*, pages 408–413. Springer-Verlag, 1989.

[Quisquater and Girault, 1989] J. J. Quisquater and M. Girault. $2n$-Bit Hash Functions Using n-Bit Symmetric Block Cipher Algorithms. In *Abstracts of EUROCRYPT '89*, 1989.

[Rabin, 1978] M. O. Rabin. Digitalized Signatures. In R. A. Demillo, D. P. Dobkin, A. K. Jones, and R. J. Lipton, editors, *Foundations of Secure Computation*, pages 155–166, New York, 1978. Academic Press.

[Rivest *et al.*, 1978] R. L. Rivest, A. Shamir, and L. Adleman. A Method for Obtaining Digital Signatures and Public-Key Cryptosystems. *Communications of the ACM*, 21(2):120–126, 1978.

[Rivest, 1990] R. L. Rivest. The MD4 Message Digest Algorithm. In *Abstracts of CRYPTO '90*, pages 281–291, 1990.

[Rompel, 1990] J. Rompel. One-way Functions are Necessary and Sufficient for Secure Signatures. In *the 22nd ACM Symposium on Theory of Computing*, pages 387–394, 1990.

[Rueppel, 1990] R. A. Rueppel. On the Security of Schnorr's Pseudo Random Generator. In *Advances in Cryptology - EUROCRYPT '89*, volume 434 of *Lecture Notes in Computer Science*, pages 423–428. Springer-Verlag, 1990.

[Sadeghiyan and Pieprzyk, 1991a] B. Sadeghiyan and J. Pieprzyk. A Construction for One Way Hash Functions and Pseudorandom Bit Generators. In *Advances in Cryptology - EUROCRYPT '91*, volume 547 of *Lecture Notes in Computer Science*, pages 431–445. Springer-Verlag, 1991.

[Sadeghiyan and Pieprzyk, 1991b] B. Sadeghiyan and J. Pieprzyk. On Necessary and Sufficient Conditions for the Construction of Super Pseudorandom Permutations. In *Abstracts of ASIACRYPT '91*, pages 117–123, 1991.

[Sadeghiyan and Pieprzyk, 1992] B. Sadeghiyan and J. Pieprzyk. A Construction for Super Pseudorandom Permutations from A Single Pseudorandom Function. In *Abstracts of EUROCRYPT '92*, pages 247–256, 1992.

[Sadeghiyan *et al.*, 1991] B. Sadeghiyan, Y. Zheng, and J. Pieprzyk. How to Construct a Family of Strong One Way Permutations. In *Abstracts of ASIACRYPT '91*, pages 55–59, 1991.

[Sadeghiyan, 1991] B. Sadeghiyan. An Overview of Secure Electronic Mail. Technical Report CS 91/1, Department of Computer Science, University College, The University of New South Wales, 1991.

[Scherift and Shamir, 1990] A. Scherift and A. Shamir. Discrete logarithm is very discreet. In *Proceedings of the ACM Symposium on Theory of Computing*, pages 405–415, 1990.

[Schnorr, 1988] C. P. Schnorr. On the Construction of Random Number Generators and Random Function Generators. In *Advances in Cryptology - EUROCRYPT '88*, volume 330 of *Lecture Notes in Computer Science*, pages 225–232. Springer-Verlag, 1988.

[Schnorr, 1991] C. P. Schnorr. FFT-Hashing, An Efficient Cryptographic Hash Function, 1991. presented at the rump session of Crypto '91.

[Schnorr, 1992] C. P. Schnorr. FFT-Hash II, Efficient Cryptographic Hashing. In *Abstracts of Eurocrypt '92*, pages 41–51, 1992.

[Seberry and Pieprzyk, 1989] J. Seberry and J. Pieprzyk. *Cryptography, An Introduction to Computer Security.* Prentice Hall, 1989.

[Shannon, 1949a] C. E. Shannon. Communication Theory of Secrecy Systems. *The Bell System Technical Journal*, 28(4):656–715, 1949.

[Shannon, 1949b] C. E. Shannon. *The Mathematical Theory of Communication.* The University of Illinois Press, Urbana, 1949.

[Shimizu and Miyaguchi, 1987] A. Shimizu and S. Miyaguchi. Fast Data Encipherment Algorithm FEAL. In *Advances in Cryptology - EUROCRYPT '87*, volume 304 of *Lecture Notes in Computer Science*, pages 267–278. Springer-Verlag, 1987.

[Silverman, 1991] R. D. Silverman. Massively distributed computing and factoring large integers. *Communication of ACM*, 34(11):95–103, 1991.

[Vandery, 1992] S. Vandery. FFT-Hash II is not yet collision-free. In *Rump Session, CRYPTO'92*, 1992.

[Vazirani and Vazirani, 1984] U. V. Vazirani and V. V. Vazirani. Efficient and Secure Pseudo-random Number Generation. In *Proceedings of the IEEE Symposium on Foundations of Computer Science*, pages 458–463, 1984.

[Webster and Tavares, 1985] A. F. Webster and S. E. Tavares. On the Design of S-boxes. In *Advances in Cryptology - CRYPTO '85*, Lecture Notes in Computer Science, pages 523–534. Springer-Verlag, 1985.

[Wegman and Carter, 1981] M. N. Wegman and J. L. Carter. New Hash Functions and Their Use in Authentication and Set Equality. *Journal of Computer and System Sciences*, 22:265–279, 1981.

[Winternitz, 1983] R. S. Winternitz. Producing a One-way Hash Function from DES. In *Advances in Cryptology - CRYPTO '83*, pages 203–207. Plenum Publishing Corporation, 1983.

[Wolfram, 1986] S. Wolfram. *Theory and Application of Cellular Automata.* World Scientific, 1986.

[Yao, 1982] A. C. Yao. Theory and Applications of Trapdoor Functions. In *the 23rd IEEE Symposium on the Foundations of Computer Science*, pages 80–91, 1982.

[Yuval, 1979] G. Yuval. How To Swindle Rabin. *Cryptologia*, 3:187–189, July 1979.

[Zheng *et al.*, 1990a] Y. Zheng, T. Matsumoto, and H. Imai. Duality between Two Cryptographic Primitives. In *the 8-th International Conference on Applied Algebra, Algebraic Algorithms and Error Correcting Codes*, page 15, 1990.

[Zheng *et al.*, 1990b] Y. Zheng, T. Matsumoto, and H. Imai. Structural Properties of One-way Hash Functions. In *Advances in Cryptology - CRYPTO '90*, pages 263–280, 1990.

[Zheng *et al.*, 1990c] Y. Zheng, T. Matsumoto, and H. Imai. Impossibility and Optimality Results on Constructing Pseudorandom Permutations. In *Advances in Cryptology - EUROCRYPT '89*, volume 434 of *Lecture Notes in Computer Science*, pages 412–422. Springer-Verlag, 1990.

[Zheng *et al.*, 1990d] Y. Zheng, T. Matsumoto, and H. Imai. On the Construction of Block Ciphers Provably Secure and Not Relying on any Unproved Hypotheses. In *Advances in Cryptology - CRYPTO '89*, volume 435 of *Lecture Notes in Computer Science*, pages 461–480. Springer-Verlag, 1990.

[Zheng *et al.*, 1992] Y. Zheng, J. Pieprzyk, and J. Seberry. HAVAL - a one-way hashing algorithm with variable lenght of output. In *Abstracts of AUSCRYPT'92*, Gold Coast, December 1992, pages 3.1-3.10.

[Zheng, 1990] Yuliang Zheng. *Principles for Designing Secure Block Ciphers and One-Way Hash Functions.* PhD thesis, Division of Electrical and Computer Engineering, Yokohama National University, 1990.

Index

Berson 54
Biham and Shamir 33
Blum and Micali 133
Blum-Micali PBG 145
Blum-Micali pseudorandom bit generator 141, 164
Brown 147

CCITT standards 36
CFB 26
CFHF 133
CIH 133
Camion and Patarin 39
Carter and Wegman 138, 159
Cellhash 41
Coppersmith 28, 31, 37, 52, 53

D-distinguishing oracle circuit 81
DES key collisions 54
DES-like permutation 66, 92
DES-type cryptosystem 116
DES 4, 14, 30
Damgard's design principle 25
Damgard's method 154
Damgard's squaring scheme 38
Damgard 32, 38, 39, 45, 133, 151
Davida 10
Davies and Price 36
Davies 9, 29
De Santis and Yung's scheme 139
De Santis and Yung 139
Deamen, Govaerts and Vandewalle 41

Denning 10
Diffie and Hellman 5, 6, 7

FEAL 4, 30
FFT hashing scheme 44
FFT-hash II 45
Feistel transformation 92
Feistel type permutation 57, 66

Girault 36, 52
Goldreich and Levin 142, 157, 163
Goldreich, Goldwasser and Micali 64
Goldreich-Levin method 165
Goldwasser, Micali and Rivest 38

HAVAL 42

ISO 13
Impagliazzo and Naor's scheme 39

Jueneman 19, 37, 50

Kam and Davida 146

L-R randomizer 107, 112
LOKI 4, 14, 30
Lai 56
Luby and Rackoff 57, 59, 66, 69, 73, 77, 105, 107
Lucifer 4

MAC 22
MD4 42

MD5 42, 54
MDC2 34
MDC4 34
MDC 23
Manipulation Detection Code 23
Massey 56
Matyas 29
Merkle and Hellman 7
Merkle's meta method 25, 45
Merkle 7, 8, 20, 31, 43, 54
Message Authentication Code 22
Meyer and Matyas 27
Meyer 29
Miyaguchi, Ohta, and Iwata 53
Moore 10

N-hash 32
Naor and Yung 133, 137
Nishimura and Sibuya 51

OFB 26
Ohnishi 70, 89
Oseas 29

PBG 133
Patarin 89
Pieprzyk and Sadeghiyan 105, 116
Pieprzyk 71, 77

Quisquater and Delescaille 28
Quisquater and Girault 30

RSA algorithm 35
RSA cryptosystem 8
RSA encryption function 145
Rabin encryption function 145
Rabin's scheme 25, 28
Rabin 25, 49
Random Matrix Hashing Algorithm

43
Rivest, Shamir and Adleman 7
Rivest 42
Rompel's scheme 140
Rompel 23, 133
Rueppel 69

Scherift and Shamir 158
Schnorr 44
Shannon 4, 57
Silverman 9
Snefru 43, 54

Toeplitz matrix 161

UOWHF 133, 137

Vazirani and Vazirani 163

Webster and Tavares 146
Winternitz' construction 29, 31
Winternitz 29, 34
Wolfram's pseudorandom bit generator 40

Yao 59, 61, 133
Yuval 25, 49

ZMI method 154
ZMI scheme 141
Zheng, Matsumoto and Imai 71, 72, 89, 92, 96, 133, 140
Zheng, Pieprzyk, and Seberry 42

active wiretapping 2
alternating group 116
authentication 2, 18
authenticity 5, 18

birthday attack 48, 49
black box test 73

cellular automaton 40
chosen plaintext/ciphertext attack 56
cipher block chaining mode 26
cipher feedback mode 26
classical message source 58
classical pseudorandom generators 58
claw-free permutation 38
collision accessibility property 137
collision free hash function 133, 136
collision intractable hash function 133
collision-free hash functions 20
collision-free hash function 38
collision-pair finder 136
complete transformation 146
correcting block attack 48
correcting last block attack 53

differential cryptanalysis 33, 54
digital signature 5
discrete Fourier transform 44
distinguisher 59
distinguishing circuit family 63
distinguishing circuits 63
distinguishing circuit 60
distinguishing probability 79

family of strong permutations 164

general attacks 49
generalized meet-in-the-middle attack 52

hard bit of a one-way function 141
hard bits 133, 141
hard-core predicates 133
hashing based on squaring 36

hashing scheme 11
hiding permutation 166

independent permutations 81
independent permutation 85
indexing 62
indistinguishability test 144
indistinguishability 60
integrity 22
inverse gate 79
inverse oracle gates 74

key collision search 30
knapsack problem 39

matrix hashing 43
meet-in-the-middle attack 28, 48, 51
message authentication 2
meta method 31

next bit test 61, 143
non-block-cipher-based hash algorithms 34
normal gate 79
normal oracle gates 74
one-time pad 6
one-way function 23, 29, 62, 142
oracle circuit 63
output feedback mode 26

parallel method 46
perfect permutation 146
perfect randomizer 112
polynomial time evaluation 62
polynomial-time test 61
polynomially samplable ensemble 135
privacy 2, 5
private-key cryptosystems 3
probabilistic Boolean circuit 59

probabilistic polynomial time algorithm
  59
probability ensemble 135
pseudorandom bit generator 60, 62,
  133, 145
pseudorandom ensemble 135
pseudorandom function generator 62,
  63
public-key cryptosystems 7

randomizer 106
redundancy 22

secrecy 5
secure hash scheme 23
secure hashing algorithm 19
semi-weak keys 53
serial method 45
simultaneous hard bits 144, 158
single-key cryptosystems 3
special attacks 49
statistical test 135
strong one-way hash function 20
strong one-way permutation 164
strong permutation 150
strongly *universal_r* family 137
substitution-permutation networks 4
super-distinguishing family 78
super-pseudorandom permutation gen-
    erator 79, 117
super-pseudorandomness 74
super-pseudorandom 85

the public-key distribution system 7
tree approach 46
two-way complete 146
type-1 transformation 94
type-2 transformation 94
type-3 transformation 95

uniform ensemble 135
universal one way hash function 133
universal one-way hash functions 20
universal one-way hash function 136
unpredictability test 144
user authentication 2

weak keys 30, 53
weak one-way hash function 21

Lecture Notes in Computer Science

For information about Vols. 1–680
please contact your bookseller or Springer-Verlag

Vol. 681: H. Wansing, The Logic of Information Structures. IX, 163 pages. 1993. (Subseries LNAI).

Vol. 682: B. Bouchon-Meunier, L. Valverde, R. R. Yager (Eds.), IPMU '92 – Advanced Methods in Artificial Intelligence. Proceedings, 1992. IX, 367 pages. 1993.

Vol. 683: G.J. Milne, L. Pierre (Eds.), Correct Hardware Design and Verification Methods. Proceedings, 1993. VIII, 270 Pages. 1993.

Vol. 684: A. Apostolico, M. Crochemore, Z. Galil, U. Manber (Eds.), Combinatorial Pattern Matching. Proceedings, 1993. VIII, 265 pages. 1993.

Vol. 685: C. Rolland, F. Bodart, C. Cauvet (Eds.), Advanced Information Systems Engineering. Proceedings, 1993. XI, 650 pages. 1993.

Vol. 686: J. Mira, J. Cabestany, A. Prieto (Eds.), New Trends in Neural Computation. Proceedings, 1993. XVII, 746 pages. 1993.

Vol. 687: H. H. Barrett, A. F. Gmitro (Eds.), Information Processing in Medical Imaging. Proceedings, 1993. XVI, 567 pages. 1993.

Vol. 688: M. Gauthier (Ed.), Ada-Europe '93. Proceedings, 1993. VIII, 353 pages. 1993.

Vol. 689: J. Komorowski, Z. W. Ras (Eds.), Methodologies for Intelligent Systems. Proceedings, 1993. XI, 653 pages. 1993. (Subseries LNAI).

Vol. 690: C. Kirchner (Ed.), Rewriting Techniques and Applications. Proceedings, 1993. XI, 488 pages. 1993.

Vol. 691: M. Ajmone Marsan (Ed.), Application and Theory of Petri Nets 1993. Proceedings, 1993. IX, 591 pages. 1993.

Vol. 692: D. Abel, B.C. Ooi (Eds.), Advances in Spatial Databases. Proceedings, 1993. XIII, 529 pages. 1993.

Vol. 693: P. E. Lauer (Ed.), Functional Programming, Concurrency, Simulation and Automated Reasoning. Proceedings, 1991/1992. XI, 398 pages. 1993.

Vol. 694: A. Bode, M. Reeve, G. Wolf (Eds.), PARLE '93. Parallel Architectures and Languages Europe. Proceedings, 1993. XVII, 770 pages. 1993.

Vol. 695: E. P. Klement, W. Slany (Eds.), Fuzzy Logic in Artificial Intelligence. Proceedings, 1993. VIII, 192 pages. 1993. (Subseries LNAI).

Vol. 696: M. Worboys, A. F. Grundy (Eds.), Advances in Databases. Proceedings, 1993. X, 276 pages. 1993.

Vol. 697: C. Courcoubetis (Ed.), Computer Aided Verification. Proceedings, 1993. IX, 504 pages. 1993.

Vol. 698: A. Voronkov (Ed.), Logic Programming and Automated Reasoning. Proceedings, 1993. XIII, 386 pages. 1993. (Subseries LNAI).

Vol. 699: G. W. Mineau, B. Moulin, J. F. Sowa (Ed Conceptual Graphs for Knowledge Representation. P ceedings, 1993. IX, 451 pages. 1993. (Subseries LNAI

Vol. 700: A. Lingas, R. Karlsson, S. Carlsson (Eds.), A tomata, Languages and Programming. Proceedings, 19 XII, 697 pages. 1993.

Vol. 701: P. Atzeni (Ed.), LOGIDATA+: Deducti Databases with Complex Objects. VIII, 273 pages. 199

Vol. 702: E. Börger, G. Jäger, H. Kleine Büning, S. M tini, M. M. Richter (Eds.), Computer Science Logic. P ceedings, 1992. VIII, 439 pages. 1993.

Vol. 703: M. de Berg, Ray Shooting, Depth Orders a Hidden Surface Removal. X, 201 pages. 1993.

Vol. 704: F. N. Paulisch, The Design of an Extendi Graph Editor. XV, 184 pages. 1993.

Vol. 705: H. Grünbacher, R. W. Hartenstein (Eds.), Fie Programmable Gate Arrays. Proceedings, 1992. VIII, 2 pages. 1993.

Vol. 706: H. D. Rombach, V. R. Basili, R. W. Selby (Ed Experimental Software Engineering Issues. Proceedin 1992. XVIII, 261 pages. 1993.

Vol. 707: O. M. Nierstrasz (Ed.), ECOOP '93 – Obje Oriented Programming. Proceedings, 1993. XI, 531 pag 1993.

Vol. 708: C. Laugier (Ed.), Geometric Reasoning for P ception and Action. Proceedings, 1991. VIII, 281 pag 1993.

Vol. 709: F. Dehne, J.-R. Sack, N. Santoro, S. Whitesi (Eds.), Algorithms and Data Structures. Proceedings, 19 XII, 634 pages. 1993.

Vol. 710: Z. Ésik (Ed.), Fundamentals of Computati Theory. Proceedings, 1993. IX, 471 pages. 1993.

Vol. 711: A. M. Borzyszkowski, S. Sokołowski (Ed Mathematical Foundations of Computer Science 1993. P ceedings, 1993. XIII, 782 pages. 1993.

Vol. 712: P. V. Rangan (Ed.), Network and Operating S tem Support for Digital Audio and Video. Proceedin 1992. X, 416 pages. 1993.

Vol. 713: G. Gottlob, A. Leitsch, D. Mundici (Eds.), Co putational Logic and Proof Theory. Proceedings, 1993. 348 pages. 1993.

Vol. 714: M. Bruynooghe, J. Penjam (Eds.), Programm Language Implementation and Logic Programming. P ceedings, 1993. XI, 421 pages. 1993.

Vol. 715: E. Best (Ed.), CONCUR'93. Proceedings, 19 IX, 541 pages. 1993.

Vol. 716: A. U. Frank, I. Campari (Eds.), Spatial Inforr tion Theory. Proceedings, 1993. XI, 478 pages. 1993.

Vol. 717: I. Sommerville, M. Paul (Eds.), Software Engineering – ESEC '93. Proceedings, 1993. XII, 516 pages. 1993.

Vol. 718: J. Seberry, Y. Zheng (Eds.), Advances in Cryptology – AUSCRYPT '92. Proceedings, 1992. XIII, 543 pages. 1993.

Vol. 719: D. Chetverikov, W.G. Kropatsch (Eds.), Computer Analysis of Images and Patterns. Proceedings, 1993. XVI, 857 pages. 1993.

Vol. 720: V.Mařík, J. Lažanský, R.R.Wagner (Eds.), Database and Expert Systems Applications. Proceedings, 1993. XV, 768 pages. 1993.

Vol. 721: J. Fitch (Ed.), Design and Implementation of Symbolic Computation Systems. Proceedings, 1992. VIII, 215 pages. 1993.

Vol. 722: A. Miola (Ed.), Design and Implementation of Symbolic Computation Systems. Proceedings, 1993. XII, 384 pages. 1993.

Vol. 723: N. Aussenac, G. Boy, B. Gaines, M. Linster, J.-Ganascia, Y. Kodratoff (Eds.), Knowledge Acquisition for Knowledge-Based Systems. Proceedings, 1993. XIII, 466 pages. 1993. (Subseries LNAI).

Vol. 724: P. Cousot, M. Falaschi, G. Filè, A. Rauzy (Eds.), Static Analysis. Proceedings, 1993. IX, 283 pages. 1993.

Vol. 725: A. Schiper (Ed.), Distributed Algorithms. Proceedings, 1993. VIII, 325 pages. 1993.

Vol. 726: T. Lengauer (Ed.), Algorithms – ESA '93. Proceedings, 1993. IX, 419 pages. 1993

Vol. 727: M. Filgueiras, L. Damas (Eds.), Progress in Artificial Intelligence. Proceedings, 1993. X, 362 pages. 1993. (Subseries LNAI).

Vol. 728: P. Torasso (Ed.), Advances in Artificial Intelligence. Proceedings, 1993. XI, 336 pages. 1993. (Subseries LNAI).

Vol. 729: L. Donatiello, R. Nelson (Eds.), Performance Evaluation of Computer and Communication Systems. Proceedings, 1993. VIII, 675 pages. 1993.

Vol. 730: D. B. Lomet (Ed.), Foundations of Data Organization and Algorithms. Proceedings, 1993. XII, 412 pages. 1993.

Vol. 731: A. Schill (Ed.), DCE – The OSF Distributed Computing Environment. Proceedings, 1993. VIII, 285 pages. 1993.

Vol. 732: A. Bode, M. Dal Cin (Eds.), Parallel Computer Architectures. IX, 311 pages. 1993.

Vol. 733: Th. Grechenig, M. Tscheligi (Eds.), Human Computer Interaction. Proceedings, 1993. XIV, 450 pages. 1993.

Vol. 734: J. Volkert (Ed.), Parallel Computation. Proceedings, 1993. VIII, 248 pages. 1993.

Vol. 735: D. Bjørner, M. Broy, I. V. Pottosin (Eds.), Formal Methods in Programming and Their Applications. Proceedings, 1993. IX, 434 pages. 1993.

Vol. 736: R. L. Grossman, A. Nerode, A. P. Ravn, H. Rischel (Eds.), Hybrid Systems. VIII, 474 pages. 1993.

Vol. 737: J. Calmet, J. A. Campbell (Eds.), Artificial Intelligence and Symbolic Mathematical Computing. Proceedings, 1992. VIII, 305 pages. 1993.

Vol. 738: M. Weber, M. Simons, Ch. Lafontaine, The Generic Development Language Deva. XI, 246 pages. 1993.

Vol. 739: H. Imai, R. L. Rivest, T. Matsumoto (Eds.), Advances in Cryptology – ASIACRYPT '91. X, 499 pages. 1993.

Vol. 740: E. F. Brickell (Ed.), Advances in Cryptology – CRYPTO '92. Proceedings, 1992. X, 593 pages. 1993.

Vol. 741: B. Preneel, R. Govaerts, J. Vandewalle (Eds.), Computer Security and Industrial Cryptography. Proceedings, 1991. VIII, 275 pages. 1993.

Vol. 742: S. Nishio, A. Yonezawa (Eds.), Object Technologies for Advanced Software. Proceedings, 1993. X, 543 pages. 1993.

Vol. 743: S. Doshita, K. Furukawa, K. P. Jantke, T. Nishida (Eds.), Algorithmic Learning Theory. Proceedings, 1992. X, 260 pages. 1993. (Subseries LNAI)

Vol. 744: K. P. Jantke, T. Yokomori, S. Kobayashi, E. Tomita (Eds.), Algorithmic Learning Theory. Proceedings, 1993. XI, 423 pages. 1993. (Subseries LNAI)

Vol. 745: V. Roberto (Ed.), Intelligent Perceptual Systems. VIII, 378 pages. 1993. (Subseries LNAI)

Vol. 746: A. S. Tanguiane, Artificial Perception and Music Recognition. XV, 210 pages. 1993. (Subseries LNAI).

Vol. 747: M. Clarke, R. Kruse, S. Moral (Eds.), Symbolic and Quantitative Approaches to Reasoning and Uncertainty. Proceedings, 1993. X, 390 pages. 1993.

Vol. 748: R. H. Halstead Jr., T. Ito (Eds.), Parallel Symbolic Computing: Languages, Systems, and Applications. Proceedings, 1992. X, 419 pages. 1993.

Vol. 749: P. A. Fritzson (Ed.), Automated and Algorithmic Debugging. Proceedings, 1993. VIII, 369 pages. 1993.

Vol. 750: J. L. Díaz-Herrera (Ed.), Software Engineering Education. Proceedings, 1994. XII, 601 pages. 1994.

Vol. 751: B. Jähne, Spatio-Temporal Image Processing. XII, 208 pages. 1993.

Vol. 752: T. W. Finin, C. K. Nicholas, Y. Yesha (Eds.), Information and Knowledge Management. Proceedings, 1992. VII, 142 pages. 1993.

Vol. 753: L. J. Bass, J. Gornostaev, C. Unger (Eds.), Human-Computer Interaction. Proceedings, 1993. X, 388 pages. 1993.

Vol. 754: H. D. Pfeiffer, T. E. Nagle (Eds.), Conceptual Structures: Theory and Implementation. Proceedings, 1992. IX, 327 pages. 1993. (Subseries LNAI).

Vol. 755: B. Möller, H. Partsch, S. Schuman (Eds.), Formal Program Development. Proceedings. VII, 371 pages. 1993.

Vol. 756: J. Pieprzyk, B. Sadeghiyan, Design of Hashing Algorithms. XV, 194 pages. 1993.

Vol. 758: M. Teillaud, Towards Dynamic Randomized Algorithms in Computational Geometry. IX, 157 pages. 1993.

Vol. 760: S. Ceri, K. Tanaka, S. Tsur (Eds.), Deductive and Object-Oriented Databases. Proceedings, 1993. XII, 488 pages. 1993.

Vol. 761: R. Shyamasundar (Ed.), Foundations of Software Technology and Theoretical Computer Science. Proceedings, 1993. XIV, 456 pages. 1993.